The Best American Science and Nature Writing 2013

The Best American Science and Nature Writing™ 2013

Edited and with an Introduction
by Siddhartha Mukherjee

Tim Folger, Series Editor

A Mariner Original
HOUGHTON MIFFLIN HARCOURT
BOSTON • NEW YORK 2013

www.hmhbooks.com

ISSN 1530-1508
ISBN 978-0-544-00343-9

Printed in the United States of America
DOC 10 9 8 7 6 5 4 3 2 1

Contents

Contents

Foreword

ON AN AUTUMN night 404 years ago, a struggling mathematics professor at the University of Padua aimed a crude, three-foot-long wood and leather telescope at the night sky. Galileo Galilei didn't invent the telescope; a trio of Dutch eyeglass makers had done that a year before Galileo used one to observe the sky over Tuscany. Nor was he the first to turn a telescope toward the heavens—an Englishman named Thomas Harriot, who sketched the moon in July 1609, claimed that honor. But Galileo was the first to truly *see* what the heavens held.

Unlike Harriot, whose drawings depicted the moon as two-dimensional, Galileo, who had had formal training as an artist, understood that the moon was not flat and smooth. By carefully studying the nightly changes in light and shadow on the lunar surface, Galileo realized that the moon had valleys, mountains, and plains not wholly unlike those on Earth. His remarkable sketches of the moon, published in 1610 in a pamphlet called *Sidereus Nuncius,* or "Starry Messenger," showed a full-fledged world, pocked and imperfect. It was just one of many observations that contradicted centuries of dogma, which held the heavens to be unchanging and perfect, composed of a rarefied celestial element, quintessence, with the planets and stars embedded on a series of concentric invisible spheres. The booklet sold out within months of publication. Its forty pages also recounted Galileo's discovery that Jupiter had four moons of its own, which completely overturned the notion that every object in the heavens orbited Earth.

And the Milky Way was not the diffuse cloud it appeared to be but rather, in Galileo's words, "a congeries of innumerable stars."

Not everyone welcomed—or even recognized—Galileo's revelations. One critic, an astrologer named Francesco Sizzi, argued that the moons of Jupiter didn't really exist, because "the satellites are invisible to the naked eye." Even some who looked through telescopes and saw the moon's imperfect surface refused to discard their old beliefs. Lodovico delle Colombe, a Florentine philosopher, argued that a perfect invisible sphere surrounded the moon. Galileo countered that perhaps the invisible sphere had invisible mountains as well.

The failure to acknowledge or understand the discoveries of science was not unique to Galileo's time. We have our own Sizzis and Delle Colombes today: politicians who deny the existence of global warming, even as glaciers shrink in Greenland and ice disappears from the Arctic (see Keith Gessen's "Polar Express" for more on the state of the Arctic), and advocates of creationism, who would see pseudoscience taught in the nation's schools, 164 years after the publication of Darwin's *On the Origin of Species*. Fortunately, we have many gifted writers—call them starry messengers—who gracefully communicate the most important stories of our time. Some of those writers, like Siddhartha Mukherjee, our guest editor, are scientists themselves, and they give us all an opportunity to peer through a lens, as it were, and see the world as it is, and not as we believe it to be.

When we do gaze through that lens, we often find a world far more beautiful and strange than anything dreamt of by medieval philosophers. Surely not even Galileo could have anticipated the paradoxes of quantum mechanics. In "Beyond the Quantum Horizon," two eminent physicists discuss, among other things, the meaning of the most confoundingly weird theory in the history of science. Nearly a century after the birth of quantum mechanics, physicists still can't agree on what it says about the nature of reality.

What would Galileo find most astounding about our time? Perhaps it would be the discovery of the enormous scale of the universe, and that it is expanding, which would have been problematic indeed for all those crystalline spheres that were once thought to hold the stars and planets. (See Alan Lightman's "Our Place in the Universe" for a lively history of the evolution of our understanding of the size of the universe.) No doubt he would marvel at our ability to live beneath the sea for days on end, as the pioneering oceanographer Sylvia Earle has. In "The Sweet Spot in Time," she shares a lifetime of insights gleaned from more than 7,000 hours underwater—nearly a year altogether. Despite all that we have inflicted on the oceans, her story in these pages shows that she still thrills to the sheer wonder of our existence on a planet unlike any other in the cosmos.

I hope this short overview whets your appetite. Siddhartha Mukherjee has selected twenty-seven stories for this collection. They're all compelling and wonderful, but I think his lyrical introduction is itself worth the price of admission. You won't soon forget it. So if I were you, I'd skip ahead right now and be prepared, as some of Galileo's contemporaries were not, to behold a field of stars.

I hope too that readers, writers, and editors will nominate their favorite articles for next year's anthology at http://timfolger.net /forums. The criteria for submissions and deadlines, and the address to which entries should be sent, can be found in the "news and announcements" forum on my website. Once again this year I'm offering an incentive to enlist readers to scour the nation in search of good science and nature writing: send me an article that I haven't found, and if the article makes it into the anthology, I'll mail you a free copy of next year's edition. Perhaps I'll manage to cajole Dr. Mukherjee into signing some copies. I also encourage readers to use the forums to leave feedback about the collection and to discuss all things scientific. The best way for publications to guarantee that their articles are considered for inclusion in the anthology is to place me on their subscription list, using the address posted in the "news and announcements" forum. Bribes and other inducements are, I'm afraid, frowned upon.

One of the pleasures of my involvement with this anthology is the opportunity to work with today's best writers. Dr. Mukherjee's book *The Emperor of All Maladies: A Biography of Cancer* should be at the top of many reading lists. Once again this year I'm indebted to Ashley Gilliam at Houghton Mifflin Harcourt. This year's anthology is dedicated to the memory of my mother, Veronica. And as always, I'm grateful to my beauteous muse, Anne Nolan, and to her pals, M. and WFB.

TIM FOLGER

Introduction: On Tenderness

In the summer of 2012, I traveled to Brno, in the Czech Republic, to visit the monastery of Gregor Mendel. I knew the barest details of Mendel's life—enough to generate an anatomical sketch but not much more. Originally from a farming family in Moravia, he had joined the Augustinian monastery in Brno in the 1830s. In 1864, working with peas in the garden of his monastery, he stumbled on arguably the most seminal discovery of modern biology: that hereditary information is transmitted from one generation to the next in the form of discrete particles of information—"genes."

The evening train from Vienna to Brno sliced its way through a spectacular landscape of farmlands and vineyards—one scintilla of green blending into another. Brno was a small town with an outsize train station. Formerly a major center of commerce, as the guidebook reminded me, protesting feebly, it had by now largely resigned itself to its fate as a way station between Vienna and Prague. In the lobby of the hotel, the concierge looked at me quizzically when I asked him about Mendel. Most of the other residents of the hotel were Russians attending a conference on oil manufacturing.

The next morning, I walked about a mile downhill from the hotel to the monastery. The building—St. Thomas's Abbey—is a plaster-and-concrete structure attached to the southern edge of an imposing church. It is as cold as a meat locker and as sparse as a prison. A faded poster of Mendel smiling mysteriously, like a rotund Mona Lisa, hangs on the edge of the boundary walls.

The walled garden in front of the abbey was overgrown and

empty. The glass hothouse, where Mendel had artificially pollinated flowers with tiny forceps and a paintbrush, had been dismantled several years earlier. The rectangular plot of land next to the building—a 12-by-6-foot mini-garden where Mendel had grown his peas for his famous experiment—was now planted, incongruously, with rows of red and white gardenias.

An auburn-haired woman was at the front desk.

I told her that I had traveled from New York to Brno to visit Mendel's monastery. "I am a geneticist," I explained, and this was a pilgrimage of sorts. Might I visit the interior of the abbey? Were Mendel's notebooks kept inside? Could I visit the room where he had tabulated his first pea hybrids or the library where he had encountered a copy of Darwin's *On the Origin of Species*?

She looked unconvinced. Apparently the abbey was closed that day. "To enter, you must send in an application," she said in Czech, and then in halting English that I could barely understand. "In duplicate."

"But I am in town for only a single night. I'm sorry, I had no idea about the application," I pleaded.

She shrugged her shoulders. "You must send in an application," she said again, with an air of finality.

My desperation was mounting. "To whom must I apply?" I asked.

"To me," she said.

I scrutinized her face. If there was even the faintest glimmer of irony, I had missed it. Well, two could play this game, I thought.

"In that case, I am applying to you now," I said. "I hereby present my application to visit Gregor Mendel's monastery." I restrained myself from executing a small bow.

The woman considered the impasse carefully. A moment of understanding passed between us, like a tiny, malevolent bolt of electricity. She looked defeated.

"No photographs, okay?" she said. She pulled out a large key from under her desk and escorted me in.

The walls inside were damp. The one-room cells that had housed the monks were largely bare, save a bed and a wooden desk. The library had about two hundred leather-bound books and a reading chair; scanning it quickly, I found nothing on botany or biology, and certainly no copy of Darwin. Mendel's room, above the refectory, was also bare, with a bed and a chair in the corner. A single evocative moment passed quickly: the wind blew the window open, and, for a second, the room became a microscope or an observatory, revealing a direct view of the rectangular garden plot below.

We made our way down a broad staircase and past the refectory. Downstairs, an inner garden was meant to grow hawkweed—another of Mendel's experimental plants—but was mostly colonized by a tangle of assorted weeds. There was a courtyard, an alleyway into the neighboring church, and a decaying niche for offering prayers. And then, as abruptly as it had started, the visit came to an end.

"Thank you for visiting the abbey," the woman said stiffly, ushering me out the door and locking it with the key.

Back in the lobby, I bought a booklet on Mendel and a T-shirt with his handwritten diagrams reproduced across it. His actual notebooks were housed elsewhere; to access them, I would need another application, possibly in triplicate.

I gave up. As I left the building, sensing the custodian's eyes scanning my back, I wondered whether the fuss had been worth it. As pilgrimages go, this had turned out to be a spectacular anticlimax.

On the train back to Vienna the next morning, I stewed in my seat, ruminating on how disappointing my visit had been. Perhaps I had expected too much. I had gone to Brno seeking something magical: an insight into the *soul* of the man who had revolutionized biology, a reconnaissance with my own intellectual history—a vivid teleportation into Mendel's life and times. But the experience had left me cold and uninspired. I felt duped. I had traveled 3,000 miles to the birthplace of genetics, and all I had gotten was a booklet and a T-shirt.

An hour out of Bratislava, though, my anger cooled. Perhaps the custodians of Mendel's legacy had—if unwittingly—achieved a rather accurate re-creation, or even a reenactment, of his life in the abbey. The rule-boundedness, the deference to authority, the *moral* disapproval at the smallest transgressions of discipline—that ever-so-slight shrug at my unfiled application—were all symptomatic; had Mendel himself been asked to curate a monument to his own stifling times, he could not have chosen a more seasoned actor to play its guardian.

Mendel's forty-odd-year stint at the Brno abbey was, indeed, deeply constrained by rules, habits, and limits. He began his experiments on inheritance by breeding field mice but was asked to discontinue them because forcing mice to mate was considered too risqué for a monk. He failed his training exams in science —notably in geology and biology—because he was unable to classify rocks and mammals using the elaborate traditional systems of classification. A sympathetic superior, Abbot Napp, allowed him to continue his experiments on peas in his garden plot, but Mendel was held to the abbey's strict routines and demands. In one of the few letters that survive, a stern note from his watchers instructs him to remember to wear his cap to church services. Mendel, for his part, was all too eager to comply. Far from a boundary-breaking, rule-bending enfant terrible, he was disciplined, deferential, and dull.

How on earth, then, did *this* man, in *this* place, unlock the secret of genes? Newton had his cometary intellect; Einstein was born a rebel and bred to defy convention; Feynman was the comic genius of physics, exposing his discipline's vanities like a jester in a court of fools. But Gregor Mendel? The founder of modern biology seems, in contrast, to have been born without contrast—a man of habits plodding his way among men in habits.

At least part of the answer, I think, takes us back to the monastery—to that minuscule rectangle of land by the refectory; to the walled garden; to the indelible image of a monk in wire-rimmed glasses tending plants—stooping, with paintbrush and forceps, to transfer the orange dust of pollen from the stamen of one flower to the pistil of the next. "It requires indeed some courage to undertake a labor of such far-reaching extent," Mendel wrote in his 1865 paper, describing an eight-year experiment on cross-fertilization that ultimately revealed the existence of genes. But "courage," I would argue, is the wrong word here. More than "courage," there is something else evident in that work—a quality that I can only describe as "tenderness."

It is a word not typically used to describe science or scientists. It shares roots, of course, with "tending"—a farmer's or gardener's activity—but also with "tension," the stretching of a pea tendril to incline it toward sunlight or train it on an arbor. It describes a certain intimacy between humans and nature—a nourishment that must happen before investigation can happen, the delicacy of labor that must be performed before the delicacy of its fruits can be harvested.

Mendel was, first and foremost, a gardener; his science began with tending. His genius was certainly not fueled by deep knowledge of the conventions of biology (thankfully, he failed that exam). Rather, it was his instinctual knowledge of the garden, coupled with an incisive power of observation, that brought him to question the nature of inheritance and thereby discover genes. The act of tending—the laborious cross-pollination of seedlings, the meticulous tabulation of the colors of cotyledons and the markings of wrinkles on seeds—soon led him to findings that could not be explained by the traditional understanding of inheritance. Heredity, Mendel realized, could be explained only by the passage of discrete pieces of information from parents to offspring. There had to be atoms of information—*particles* of inheritance—moving from one generation to the next. Tending generated tension—until the old fulcrum of biology was snapped in two.

When I witness science in action, I see this tenderness in abundance. On Monday mornings, the graduate students and postdoctoral researchers in my laboratory rush in to work to look at how their cells have grown over the weekend. The best of these researchers have a gardener's instinct. Some of the cultures need nourishment, they know; others, like ferns, need to be left alone to inhabit the corners of incubators; yet others must be coaxed with growth factors to flourish.

Look closely among scientists, and you find this quality everywhere. There is tenderness in the chemist measuring and remeasuring salts in the hood; in the mathematician kneading his equations to understand the shape of the cosmos; in the marine biologist learning to talk to dolphins (read Tim Zimmermann's "Talk to Me"). Newspapers may bring us news of a scientific-industrial complex that is increasingly depersonalized—algorithmic, disembodied, and run by robots. The lab is apparently a factory. Terabytes of data are churned through supercomputers to generate gigabytes of information; the scientist punches numbers into a machine and awaits revelation. But ask a real scientist, and you get a profoundly different image of how "real" science happens. In an age of increasingly mechanized production, the genesis of scientific knowledge remains an unyieldingly, obstreperously hand-hewn process. It is among the most human of our activities. Far from being subsumed by the dehumanizing effects of technology, science remains our last stand against it.

I have chosen the essays in this volume with an ear for tenderness. Most of the selected essays share a common thread: they describe how science *happens*. They don't present facts alone (although facts are abundant in them). They describe the extraordinary process by which scientists extract those facts from the grim soil, roots and tendrils intact, to glean knowledge about the inner workings of nature.

Listen, then, for tenderness in these essays. It is present, of course, in Katherine Harmon's sprawling Russian novel of a piece, "The Patient Scientist," about a prominent New York immunologist with pancreatic cancer who becomes his own experimental subject (I knew Ralph Steinman, the scientist in question, and was struck by Harmon's devastatingly honest and moving portrait of him). And it can be found equally in Jerome Groopman's "The T-Cell Army," about the once-moribund discipline of cancer immunology coming to life in the laboratory and the clinic.

It is easy to find tenderness in the remarkable essay "Autism Inc.," about the parent of an autistic child who starts a company called Specialisterne, Danish for "the specialists"—"on the theory that given the right environment, an autistic adult could not just hold down a job but also be the best person for it." It may be harder to discern tenderness in Kevin Dutton's coldly wise "The Wisdom of Psychopaths"—but it's there, roiling just beneath the surface of this story of a psychologist who seeks to understand the workings of a psychopath's mind. In talking to dozens of patients confined to a high-security psychiatric prison in England, Dutton emerges with a strangely complex understanding of what psychopathy is and how it defines its obverse: empathy.

Steve Weinberg's "The Crisis of Big Science" is a cry from the heart that is meant to provoke political action. Sometime in the next decade, Weinberg writes, physicists are going to ask their governments to fund the building of the most powerful linear accelerator ever built. This accelerator—not the Large Hadron Collider but the Even Larger Hadron Collider—will supposedly smash its way through an experimental impasse that particle physicists apparently find themselves stuck in, allowing them to prove or disprove models about the fundamental nature of matter and energy. But notably, Weinberg doesn't confuse big science with great science. His essay begins with a description of Ernest Rutherford's discovery of the atomic organization of matter. Rutherford's experimental team, Weinberg informs us, "consisted of one postdoc and one undergraduate," and was funded by a grant of £70 from the Royal Society. Rutherford worked largely alone, fussing over his instruments and detectors; he was Mendel in an atomic garden. The particle physicists of tomorrow might indeed need bigger accelerators, as Weinberg argues. But to transform big science into great science, I suspect, they will need to channel Rutherford's spirit into their much larger atomic gardens.

One set of essays describes the measurement, reconstruction, and surveillance or restoration of impossibly fragile systems (read David Owen's "The Artificial Leaf," Michael Specter's "The Deadliest Virus," David Quammen's "Out of the Wild," Mark Bowden's "The Measured Man," or Elizabeth Kolbert's "Recall of the Wild"). Robert M. Sapolsky's "Super Humanity" and Stephen Marche's "Is Facebook Making Us Lonely?" converge on a similar and deeply affecting thought—that humans may have created modern environments (including virtual environments) that are peculiarly maladapted for their intended purpose: rather than assuaging anxiety and bringing communities together, these environments provoke anxiety and encourage lonesomeness. Sapolsky's answer to this quandary is particularly potent: far from rejecting science as dehumanizing, he turns to it as a force of creative regeneration. To tend the wounds of the human psyche—to restore what has been lost—he argues, we need more science, not less.

And look for tenderness, lastly, in "Shattered Genius," a profile of the Russian mathematician Grigori Perelman, who solved the infamously thorny Poincaré's Conjecture but could not be bothered with collecting the million-dollar prize for doing so. Perelman is a purist. He despises the crassness of the world, with its academic competitions and silly prizes; he will not be put up for display like an animal in a zoo. There is something raw about him —a hothouse temperament so delicate that the world bruises it all too easily (the profile reminded me of Marianne Moore's lines on the student, who is reclusive "not because he / has no feeling, but because he has so much"). When an all-too-eager journalist hunts him down to talk about his uncollected prize, Perelman snaps at him with a sentence that a gardener might be proud of: "You are disturbing me. I am picking mushrooms."

SIDDHARTHA MUKHERJEE

*The Best American Science
and Nature Writing 2013*

False Idyll

FROM *Orion*

OF ALL THE feelings said to sweep over us in wild places—awe, peace, a sense of the divine—there are a few that rarely get mentioned. My last two-week trip into the woods, for example, was frankly depressing. The year had been a cold one, and the forest was not its usual refulgent self. A black bear was hanging around, skinny and sickly from the bad berry crop and probably bound for death by starvation in its winter den. Pink salmon had just begun to spawn in a nearby creek, where their battered bodies were a reminder of the grand cycle of life, yes, but were also an intimately dismal spectacle. Then I discovered a colony of bats, the year's pups just learning to fly. Not a lot is known about the mortality rate of bats in this fledgling period, but I am inclined to predict it is high. The little ones peeped fearfully before their maiden flights, and with good reason—I watched several crash into the tall grass, unlikely ever to make it home again. They might, at least, make easy meals for the garter snake I saw that had somehow lost half its face.

All of this took place in a valley that, blessed with steep slopes, icy winters, wet summers, and remoteness from the world's stock exchanges, has somehow retained the full complement of predators, including wolves, grizzly bears, and mountain lions. I do indeed feel awe in that place, but not much peace. By day I carry pepper spray, and by night I sleep with a twelve-gauge shotgun close at hand, because a couple of years ago a bear tried to break into my "cabin"—a ninety-year-old homestead shack that can't even keep out the rain—in the first light of dawn. If a god is in charge of the area, he is surely of the mercurial, Old Testament variety.

The idea that nature is a bittersweet and sometimes forbidding place is not, as they say, currently trending. More prevalent is the view reflected in a recent caution from the *Chicago Manual of Style* editors that capital-N "Nature" is to be used only to denote "a goddess dressed in a flowing garment and flinging fruit and flowers everywhere." The comment is tongue-in-cheek, but the point is well taken. The natural world is increasingly seen as a gentle and giving realm of the spirit. In some cases, this view is actively religious or quasireligious, whether we are speaking of the biosphere as the provident Earth Mother, the being-of-beings that is James Lovelock's Gaia, or simply the handiwork of one or another god. But above all else, the actual experience of being in nature seems to affirm its essential holiness. The natural world *feels* like a spiritual respite: a literal sanctum, where we are safe to reconnect to what is larger than ourselves. Compared to the cosmic rhythms of mountain, sea, and sky, it is ordinary daily life—driving at rush hour, punching security codes, navigating a shape-shifting digital culture—that seems hostile.

Yet there is a serious problem with our idea of sacred nature, and that is that the idol is a false one. If we experience the natural world as a place of succor and comfort, it is in large part because we have made it so. Only 20 percent of Earth's terrestrial surface is still home to all the large mammals it held five hundred years ago, and even across those refugia they are drastically reduced in abundance. The seas have lost an estimated 90 percent of their biggest fish. For decades there were almost no wolves, grizzly bears, or even bald eagles in the lower forty-eight, and modern recovery projects have brought them back to only a small fraction of their former ranges. Scientists speak of an "ecology of fear" that once guided the movements and behavior of animals that shared land- and seascapes with toothy predators—an anxiety that humans once shared. In much of what's left of the wild, that dread no longer applies even to deer or rabbits, let alone us. The sheer abundance and variety of the living world, its endless chaos of killing and starving and rutting and suffering, its routine horrors of mass death and infanticide and parasites and drought, have faded from sight and mind. We have rendered nature an easy god to worship.

If humankind's relationship to the wild were to be embodied

by just one of the gods we have invented, I would nominate Janus, the twin-faced deity of the ancient Romans. Our sense of the divine can connect us to nature, but it can divide us from it as well. Spirituality can help us see ourselves as kindred to every living and nonliving thing, all sprung from the same celestial dust. This primeval understanding remains deep and broad today, revealed everywhere from the Garden of Eden story shared in one form or another by Christians, Jews, and Muslims; to the Tibetan name for Mount Everest, Chomolungma, the Holy Mother; to $2,995 shamanic journeys of reconnection to Mother Earth in Sedona, Arizona, complete with one-night vision quests, "weather permitting." On the other hand, spirituality has long been used to place ourselves on a pedestal above the rest of creation. The Garden of Eden story includes instructions to "fill the earth and subdue it" and to "have dominion" over every living thing, among other phrases that amount to a mission statement for latter-day capitalism; Mount Everest is a challenge to be conquered; and that same Arizona wilderness retreat promises to refresh the "natural power that is your birthright."

Old Janus has been staring in these opposite directions a long time—the tension between being a part of nature and standing apart from it is elemental to what it means to be human. "The archaeological record encodes hundreds of situations in which societies were able to develop long-term sustainable relationships with their environments, and thousands of situations in which the relationships were short-lived and mutually destructive," wrote the Arizona State University anthropologist Charles Redman in his seminal 1999 book *Human Impact on Ancient Environments*. The pattern Redman points to is not, as some might suppose, divided neatly between destructive societies in the lineage of so-called Western civilization and sustainable societies in the more earth-toned traditions often associated with, for example, Native Americans. A recent scientific review of human impacts on the oceans found "overwhelming" evidence that aboriginal coastal cultures "often" depleted their local environments; in fact, the editors speculate that it may have been the struggle to survive in increasingly degraded surroundings that gave rise to the conservation values that many Native Americans appear to have held at the time of European contact. If so, then 1492 was a clash of Janusian timing: European nations reveling in the discovery of God-given riches just as

Native American cultures were formulating a spiritual understanding of natural limits.

We know which of those two worldviews prevailed in the centuries that followed—a history that astounds us with the extinction or near-extinction of even the most superabundant creatures, from the great auk to the buffalo to the Atlantic cod, though these iconic species are best thought of only as reminders of a wholesale assault on animate life that left no species unscarred. In the midst of it all, a countercurrent emerged. A small minority of people still mark the beginnings of that turning with the 1864 book *Man and Nature*, by George Perkins Marsh, a pioneer of ecological thought. With the exhausting thoroughness of autodidactic science-geekery, he presented an inventory of "the extent of the changes produced by human action in the physical conditions of the globe we inhabit." For the most part, however, Marsh is a footnote, massively overshadowed by his more lyrical, less empirical contemporaries. I don't even need to use their first names: nature writing in the tradition of Emerson and Thoreau, of Wordsworth and Coleridge, has called on us to see the face of God in every trembling leaf ever since. To do otherwise is to fall into the cold rationalism so often said to have betrayed the wild world.

This modern love of the Earth is ironic—it is a reaction against the destruction of nature but is also a product of that destruction. Witness Great Britain, once home to deep forests, bears, wolves, wild boars, wild oxen. We celebrate England's Romantic poets for seeing divinity in a landscape that others found dark and threatening. Yet the Romantics were only opening their eyes to a new reality: almost every threat posed by that wild landscape had been vanquished. By the time of the Romantics, Britain was much as it is today—a deforested island, its fauna largely reduced to butterflies, birds, and hedgehogs.

The pattern repeated itself on the American shore. Thoreau wrote from a forest that had lost its capacity to instill fear in a young man's heart. (Marsh could have detailed this history for him; Marsh's childhood home near Woodstock, Vermont, had in his lifetime lost its moose, wolves, and mountain lions, and seen its spruce and hemlock forests replaced with European trees.) Annie Dillard's pilgrimage to Tinker Creek plays out in a denuded Virginia, and even Edward Abbey, that singular voice of wildest America, went to his deathbed never having seen a free-living grizzly bear. Such versions of nature still inspire wonder—I held a wild hedgehog in my hands last year and was speechless with the thrill of it. In fact, one might argue that the works that have brought us closest to nature have *depended* on a more welcoming wilderness. But another truth should be foremost in mind: that what we call nature today is a kinder, gentler, more depauperate world than at any time since at least the late Paleozoic, some 300 million years ago. Nature is not a temple but a ruin. A beautiful ruin, but a ruin all the same.

According to recent statistics, most people on Earth now live in cities, with few if any daily reminders of things ecological. There is considerable evidence that this disconnect costs us at a personal level. Among the most durable findings in the field of environmental psychology, for example, is that we prefer natural settings over the built environment. Among natural landscapes, we show the greatest preference for open spaces dotted with trees, with a little water nearby. (Picture the views from the apartments that border Central Park in Manhattan; as the biologist E. O. Wilson puts it, "To see most clearly the manifestations of human instinct, it is useful to start with the rich.") These preferences have a consistency across cultures and generations that approaches evolutionary natural law.

I want to call attention to two aspects of these discoveries. The first is that the salient feature of our most preferred environments —savanna-like spaces—is long sightlines, which would have helped us to survive the eons when our species was still a link in the wild food chain. In other words, we prefer nature when it is unthreatening, and on that count, we have had our wish. The second point is that we nonetheless have a deeply embedded psychological attachment to the living world. Having lost our daily communion with that world, our modern spiritualization of it can be seen as a kind of prosthetic—or, if you prefer, a way of turning up the volume on a signal that is increasingly faint. We have created an imaginary connection with nature because we lack a tangible one, and we carry that connection in spirit because we no longer follow it in body. The sense of the divine that many feel in wild places is less a bond with nature than another symptom of the absence of that bond.

Ecologically speaking, this sanctified nature is not nearly enough. "We live more and more in an enchanted illusion of what nature is, which I think is counterproductive to conservation," says the Cornell University biologist Harry Greene. It's the back half of that statement— *counterproductive to conservation*—that contains surprises. At the time, Greene was responding to the movement that seeks, in effect, to protect feral mustang horses in the American West from natural life and death, permitting neither human culling nor wild predation nor starvation from drought or harsh winters, and instead using pharmaceutical contraceptives to control the population. This approach falls close to the farthest end of the spectrum of enchantment, where we find "end of suffering" activists who see a high moral calling in technocratic intervention against every cruelty that regulates natural systems: no more frogs swallowed alive by snakes, no more calf elk gored by grizzlies in front of their mothers' eyes, no more exhausted hummingbirds drowned during their arduous migration across the Gulf of Mexico. "Let's aim to be compassionate gods," concludes one essay from the end-of-suffering sect, "and replace the cruelty of Darwinian life with something better."

But such extreme examples aren't necessary. We might instead simply reflect upon the ecological consequences of our having created a wild world that has, for the most part, liberated us from fang and claw and distanced us from unseemly reality. Writing in the 2010 book *Trophic Cascades*, editors John Terborgh and James Estes, both prominent ecologists, describe the simplification of nature's architecture by human actions as a crisis "every bit as serious, universal, and urgent as climate change." When fishermen's nets fill not with fish but jellyfish; when pestilent tsetse flies spread with the scrublands once held in check by browsing elephants; when overpopulating deer eat the flower gardens of suburban America—all of these bear the markings of the ecological cascade. Of greatest concern is that most of what has changed, and how, and at what cost, has not even been calculated. Here's one example that hints at the scale of the losses: the best available estimate suggests that whales before whaling ate up nearly 65 percent of the energy—as transformed into living things—produced yearly in the world's oceans. Paradoxically, however, the same seas that teemed with ravenous whales also brimmed with other creatures great and small, from swordfish to shad to oysters. "We know very little about the direct and indirect effects of reducing whale populations by more than 90 percent, but they must be substantial," note Terborgh and Estes, with the typical restraint of lifelong scientists. It's knowledge that could be of some use to us right now. By conservative estimates, a single animal—us—now consumes at least a quarter of the annual productivity of the planet, with the critical difference that our myriad hungers are satisfied only at enormous expense to the abundance and variety of species.

Are we to blame a global society's accumulating insults against the biosphere on people who meditate in the desert or find divinity beneath the redwoods? No. But the way you see the world determines much about the world you are willing to live in, and the spiritual lens has failed us as a tool for seeing clearly. Here are Terborgh and Estes again: "There is little public awareness of impending biotic impoverishment because the drivers of collapse are the *absence* of essentially invisible processes . . . and because the ensuing transformations are slow and often subtle, involving gradual compositional changes that are beyond the powers of observation of most lay observers." Our collective response to these shifts in our surroundings, as Michael Soulé, a founding figure in conservation biology, puts it, is to "excuse, permit, and adapt." The romanticization of a denatured living world is one such adaptation. We have turned a fierce and ambiguous nature into a place of comfort, and if we embrace the result as a sanctuary of the soul, to be visited every second or third long weekend, then we may ultimately see little purpose in returning to a deeper and more risky engagement. We'll end up with the twin faces of Janus both looking the same direction, having found all the wildness we need in the tamed.

Every year, I try to return to that cabin where the bears roam and the salmon spawn and die, and the baby bats risk their new lives in fragile flight. There is no road; the access is by train, or by boat across a river of terrifying cold and current. I once told people that I went there for the peace and quiet, to escape into the sublime, and that was not entirely a lie. But I have to admit that I often feel a growing dread as the moment of entry into that wilderness approaches. It's not the solace of mountain and forest that keeps drawing me back. It is something more demanding.

Every day in that wild place is an opportunity to pass time with eagles, ravens, toads, snakes, moose, grouse, salmon, and the year's local black bear, which somehow always seems to be everywhere at all times. I often find myself filled with wonder, but the challenge of living nearer to nature will never be having to cope with more beauty, or that our hearts may explode from so much swelling. Instead, the challenge comes from the wilderness's countless mortal shocks, from maggots teeming in the brainpan of a dead deer, to the steady watchfulness required of life among large predators, to weirdly disturbing realizations such as that adult mayflies have no mouths, no digestive tracts, no anuses. Yet another memory from this past year's visit leaps to mind: a strange preponderance of bleeding-tooth fungus, *Hydnellum peckii*, which weeps transparent beads of red liquid across the white pulp of its mushroom cap. If the bleeding-tooth fungus is the answer to any question, that question could only be *"Why?"*

If the modern spiritualization of nature is the product of distance and diminishment, observations such as these are the opposite, the outcome of muddy hands and scratched skin, of having time to waste in places where our species is a curiosity and a potential source of protein. Slowly, haltingly, I am coming to see the community of species around my cabin with the same eyes with which I have come to see other communities—to the extent that even that word, "community," sounds clinical and precious to my ears. Think instead of your friendships, or your neighborhood, those fragile constructions of toleration and embrace, of the heartwarming and the bleak. We understand our friends and neighbors as imperfect, even essentially tragic, and yet, at our best, we know that they are a part of us—that we are enriched when they are enriched, impoverished when they are impoverished. I am still new to the neighborhood of salmon, cedar, and raven, and I won't claim any insight into their world that is more profound than this: I feel their absence when I leave, and it's their presence that always draws me back again.

It hasn't been my experience that full-force nature directs the mind toward thoughts of positive vibrations or divine master plans. Nature itself is enough, its stories written in blood and shit and electrons and birdsong, and in this we may ultimately find all the sacredness we seem to need.

One final story: Several years ago, I interviewed a woman named Sally Mueller, who had moved with her family from New Mexico to the remote Tatlayoko Valley of British Columbia. She had, in effect, made the decision that I have never found myself quite ready to make—to seek a life in the wilderness. There, many happy years later, she was charged by a sow grizzly protecting her cubs. The animal stopped only inches away and, roaring, swiped with a paw, slicing through two layers of clothing and the flesh of Mueller's thumb. Only then did the mother bear's fury drain away. The grizzly retreated; the scales of life and death tilted back into balance; the crawl of time returned to its regularly scheduled programming.

"It was really a highly spiritual experience for me," Mueller said. She shared that revelation cautiously, aware that it would be difficult to understand. But in those terrible instants, she said, she knew that the bear was only doing what it must, and so was she, and so, too, were even the meadow grasses and the trees, the earth and the sky, and all of it was blurred into a pattern too infinite and ancient to explain. At last, Mueller found the words for the feeling: "It was just like coming home."

BENJAMIN HALE

The Last Distinction?

FROM *Harper's Magazine*

HUMAN BEINGS HAVE long sought a definite marker between themselves and "the animals." In the 1960s, toolmaking was considered such a uniquely human behavior that when Jane Goodall witnessed chimpanzees modifying twigs to root for termites, the naturalist Louis Leakey responded, "Now we must redefine *tool*, redefine *Man*, or accept chimpanzees as human." Since then, other animals—crows, most recently—have been seen making and using tools. Ethological observation has similarly eroded other distinctions humans have claimed for themselves. But there remains a tradition—in literature and philosophy as much as in science—of treating language as the Rubicon that only humanity has crossed. In *Paradise Lost*, when Satan, disguised as the serpent, begins talking to Eve, she says in astonishment, "What may this mean? Language of man pronounced / By tongue of brute, and human sense expressed?" Animals do not talk. The idea is unnatural, satanic.

Speculation on the origin of human language was long discouraged among linguists; inquiry into the subject was formally banned by the Société de Linguistique de Paris in 1866, and the taboo thereby established persisted for nearly a century. The moratorium, a famous incident in the history of linguistics, began in the earliest days of Darwin's influence, after the publication of *On the Origin of Species* but a few years before the publication of *The Descent of Man*, in which Darwin first explicitly discussed human evolution—including the evolution of language.

Of modern history's important thinkers, Darwin may be the most chronically oversimplified. Distortions of his thinking began not long after his death. He treated humanity as a part of nature rather than over and above it, upsetting Europe's philosophical tradition of the Great Chain of Being: the hierarchical ordering of all creation, rising in increments toward man's perfection. This model of life was so firmly accepted that it survived even among those who accepted Darwin's work, leading to a widespread misunderstanding illustrated by a graphic so elegant (and so reductive) that it's become a pop-semiotic stand-in for the theory of evolution: the left-to-right single-file march of an ape morphing into a man, with its implication that evolution is a teleological progression and *Homo sapiens sapiens* the goal. The illustration does less to explain evolution than to reinforce the inaccurate (and specifically Western) idea of a radical break between humans and other animals.

Descartes, who wrote extensively on the philosophical problem of animal consciousness, argued that all nonhuman animals are instinctual automata, whereas humans alone think—*cogitant ergo sunt*—and therefore possess souls. The impulse to draw a circle around humanity underlies the question "What makes us human?" The way we phrase the question—which presupposes that the answer must be a definite *thing* we possess—tends to make language the most satisfactory answer.

Hence our fascination with feral children—the Wild Boy of Aveyron, Kaspar Hauser, Genie, and so on—cases of human beings isolated and deprived of language during the crucial early-acquisition period. What would it be like to have a consciousness but be unable to think in articulate language? For most people, to imagine the experience of inhabiting such a consciousness is close to impossible. The animal scientist Temple Grandin has written much on this subject, asserting that her autism lends her a unique insight into the way animals—cows, in her line of work—experience the world: wordlessly. "I think in pictures," she writes in the opening pages of her memoir.

> Words are like a second language to me. I translate both spoken and written words into full-color movies, complete with sound, which run like a VCR tape in my head. When somebody speaks to me, his words are instantly translated into pictures. Language-based thinkers often find this phenomenon difficult to understand, but in my job as an equipment designer for the livestock industry, visual thinking is a tremendous advantage.

It would be absurd to suggest that because Grandin does not think primarily in language she isn't conscious, but the importance of language as a distinct marker between the human and "the animal" mind is still lodged in the Western models of consciousness.

In the thirties, the psychologists Winthrop Kellogg and Luella Kellogg briefly raised a chimpanzee named Gua alongside their own infant son, Donald. They aborted the project after nine months because Donald seemed to be picking up more behaviors from the chimp than vice versa. In a longer and more involved experiment that began in 1947, another psychologist couple, Keith and Cathy Hayes, attempted to raise a newborn female chimp named Viki as a human child. After seven years of home rearing and intensive vocal training (including speech-therapy techniques such as physical manipulation of the mouth), Viki could articulate, in a breathy and almost inaudible voice, four words: "mama," "papa," "cup," and "up."

These early experiments focused on language production over comprehension. But ape anatomy does not readily allow articulations of the kind necessary to speak. The human vocal apparatus consists of the larynx, the throat, the nasal cavity, the tongue, and the lips—all of which are shaped differently in nonhuman apes. Chimps' vocal tracts are shorter and straighter than ours, with higher larynges. When humans speak, moreover, we accomplish what's called a velopharyngeal closure by briefly blocking off air to the nasal cavity with the soft palate, allowing us to articulate hard consonants. Apes do not have this capability.

Even as far back as the 1920s, scientists wondered about the possibilities of gestural communication. "I am inclined to conclude from the various evidences," wrote Robert Yerkes, an early American pioneer of primatology, "that the great apes have plenty to talk about, but no gift for the use of sounds to represent individual . . . feelings or ideas. Perhaps they can be taught to use their fingers, somewhat as does the deaf and dumb person, and thus helped to acquire a simple, nonvocal 'sign language.'" Recognizing that nonhuman apes, though physiologically unable to produce the same range of sounds as humans, often communicate gesturally, another psychologist couple, Allen and Beatrix Gardner, of the University of Nevada, Reno, began experimenting with sign language in 1966, using as their subject a female chimp named Washoe. The focus on sign language, which resolved the main problems of the Hayes experiment, was also influenced by Jane Goodall's and Adriaan Kortlandt's ethological reports that chimps in the wild have systems of gestural communication that are highly complex and cultural, varying from one social group to another.

The Gardners housed Washoe in a trailer in their backyard and enlisted a small staff of graduate students to help teach her American Sign Language. Allen Gardner, a strict experimentalist, began the study with Skinnerian conditioning techniques, which are undeniably useful in any animal training. For example, they would wait for Washoe's "hand-babbling" to form something that looked like an ASL sign, then reward her with food or such displays of approval as clapping, smiling, and tickling. They would refine the sign with further rewarding and try to condition her to use it in correct contexts. They abandoned these methods not long into the experiment because Washoe had learned only one sign: "funny." Roger Fouts, who worked closely with Washoe her entire life, writes in his memoir, *Next of Kin,* that after the first year of the experiment Washoe picked up signs almost entirely from watching humans use them.

The methodological trickiness of the Gardners' experiment plagued it and other sign-language experiments to come. ASL is a fully developed language, with movements and facial expressions that work together to create meaning; what Washoe learned was not ASL per se, but a collection of modified ASL signs that were fluid and subjective. Their interpretability became a major problem. At the outset of the experiment, the Gardners kept records of every sign—or near-sign—Washoe made, marking the time she made it, and in what context. As Washoe's vocabulary grew and she began signing more frequently, data collection became difficult; soon the experimenters would record only her use of new signs. In order for a sign to be added to her theoretical vocabulary, three independent observers had to document that she made a "spontaneous, well-formed, and appropriate use of the sign." That sign then went on the official list of Washoe's working vocabulary only when she had used it spontaneously, articulately, and appropriately every day for fifteen days. By this measure, in 1970, four years into the experiment, Washoe had an active vocabulary of 132 signs. Although the Gardners tried to be strict with their data collection, skeptical linguists accused the experimenters of interpreting their data too generously. "Spontaneous," "well-formed," and "appropriately used" are in the eyes of the beholder—even three independent ones.

How does one determine whether an ape has made an ASL sign? Neither the Gardners nor the graduate students who worked with them were fluent in the language. "Each week I attended ASL classes at the Gardners' house," writes Roger Fouts, "but most of my learning came on the job with Washoe and her other student companions." Being nonfluent, the experimenters were probably poorly equipped to teach ASL to a chimp—or anyone, for that matter. They were learning signs at the same time Washoe was, which meant they were also probably poorly equipped to interpret them. "Often the project directors themselves were uncertain about how a particular sign should be made," writes Arden Neisser about Washoe in her book on sign language and the deaf community, *The Other Side of Silence*. "By the time it was taught to the chimp, 'It lost something in translation,' said a deaf friend ruefully."

The Gardners' experiment was a direct precursor to the Columbia University psychologist Herbert Terrace's whimsically named Nim Chimpsky project, which was beset by many of the same problems, as well as new ones, mostly of Terrace's making. The most ambitious and publicly visible of several sign-language experiments with apes throughout the 1960s and '70s, Terrace's Project Nim is the subject of a 2011 documentary of the same name, directed by the Academy Award–winning filmmaker James Marsh. Following upon Elizabeth Hess's 2008 book, *Nim Chimpsky: The Chimp Who Would Be Human*, the film chronicles the chaotic life of Terrace's subject.

Noam Chomsky, the punny namesake of both the project and the chimp, is never mentioned in the film, which focuses on the human elements of Nim's story rather than the scientific controversies surrounding it. This is an odd absence, since Chomsky more than any other thinker upheld Descartes' torch of human exceptionalism in the twentieth century. Chomsky's theories of transformational grammar, universal grammar, and the innateness and human uniqueness of language defined the debate over language for decades. These theories rest on the "poverty of the stimulus" argument: language is so complex, and infants learn it in such relatively little time, that it can't possibly be learned entirely through external stimulus; there must be an innate language-acquisition "device" or "organ" in the human brain. We were told to look for it somewhere in the left hemisphere.

Prior to recent scientific advances, neural anatomy was a dark frontier, and there was no clear evidence for or against the language-acquisition device. All we had was Chomsky's promise that it must exist and that it must govern our understanding of grammar and syntax. Rather than argue with Chomsky's anthropocentric definitions, some researchers aimed to prove that an animal could communicate in ways that filled out Chomsky's checklist of what makes language—principally, that an animal could come to understand and use grammar. Proving that an animal could be "taught" to communicate using language—as narrowly conceived by Chomsky—became a holy grail for language researchers. Herbert Terrace sought this prize by way of Nim Chimpsky.

There is something glib and thoughtless about bestowing on another conscious being a pun for a name. Glibness and thoughtlessness, as one sees in the documentary, are just a couple of Terrace's winning traits, and Nim Chimpsky's name was only the first indignity in a life full of indignity and suffering, which is the main subject of Marsh's film.

Terrace, who still conducts research at Columbia, planned to raise a chimpanzee in a human home, with no contact with other chimps, and immerse him in sign language from infancy. Just days after his birth in Norman, Oklahoma, in 1973, Nim was taken from his mother—who had previously had six infants taken from her for experiments. Terrace turned him over to Stephanie LaFarge, a former student (and a former lover) of Terrace's, who had generously and perhaps recklessly volunteered her own New York home and large family to take in baby Nim as a foster child. The LaFarges were supposed to begin speaking to Nim in ASL within months, though not a single member of the household knew the language. Stephanie's husband, W.E.R., was a poet with a ponytail and a patrician pedigree; they lived with a swarm of children from previous marriages in a brownstone on Manhattan's Upper West Side; and everyone in the picture seems to have been smoking a lot of pot. Now add to this environment one infant chimpanzee.

As Stephanie's daughter, Jenny Lee, says in the film, "It was the seventies."

Herb Terrace comes across so negatively that one wonders how much thought he gave before consenting to filmed interviews. Disliked by every other interview subject, Terrace appears irresponsible, smug, careless, cowardly, disloyal, vain, and given to having sex with his students. During the experiment, Terrace had an affair with Laura-Ann Petitto, who was an eighteen-year-old undergraduate when she began working on the project and who became one of the most involved of Nim's early caregivers. Petitto worked with Nim while he was still living with the LaFarges (Stephanie LaFarge calls her, with jealousy and a sniff of classism, "a cute little thing from Ramapo"), persisting in her work as the chimp ripped apart the LaFarges' curtains, books, and marriage, and moving with him when Terrace secured for the experiment a sprawling Georgian estate in the North Bronx that was owned by Columbia and had been sitting empty for years.

Terrace's affair with Petitto is noteworthy mostly because it interfered with the project. In Petitto's words, Terrace "abruptly" ended their romantic involvement—a diplomatic phrasing that sounds like the shutting of a cellar door. Whatever happened, she left the experiment, upsetting the balance of Nim's emotional life yet again. "It's the humans I wanted to leave," says Petitto, "not the chimp."

Terrace's participation in the experiment was by all accounts fairly minimal. He showed up for photo ops, and his name was of course listed first on the resulting paper, but he left the bulk of the work to his students. Bill Tynan, another of Nim's early caregivers, describes Terrace as "an absentee landlord" who only occasionally put in an appearance at the mansion where Nim was imprisoned in luxury like a mad aristocrat out of some gothic novel.

After four years, Terrace abandoned the experiment, largely because Nim's increasingly violent and unpredictable behavior—he was, after all, a growing chimpanzee—had created insupportable liabilities. "I was probably worried that she would sue me," Terrace says—with characteristic bluntness and lack of self-awareness—of an incident in which Nim grievously injured Renee Falitz, a sign-language interpreter who was the only person fluent in ASL ever to work on the project. So Terrace called an all-hands-on-deck meeting at which he shocked and angered his staff by announcing that the experiment was now over. Terrace had Nim tranquilized and flown back to his birthplace. Nim went to sleep in his palace and woke up in a hellish place now infamous among captive-ape researchers for its inhumaneness.

"It turned out to be a surprisingly more primitive facility than I remembered," Terrace says of the compound run by the University of Oklahoma psychologist and animal breeder William Lemmon, which served as a sort of chimpanzee-research hub in the 1960s and '70s. Electric fencing, metal cages, guns, and cattle prods were Lemmon's tools of subjugation; this was a place where chimps were treated like animals—or, rather, prisoners—not like spoiled human children. (In his memoir, Fouts claims that Lemmon wore a ruby ring that he had trained his chimps to kiss. But that's another story.) Nim entered this environment having never met a chimpanzee other than, however briefly, his mother. In the most affecting moment of Marsh's film, Terrace visits Oklahoma a year after leaving Nim there. Nim recognizes Terrace and erupts with jubilant relief, shrieking and rushing to hug him. Bob Ingersoll, who worked at Lemmon's facility, infers Nim's thoughts on seeing Terrace again: "Holy shit! I'm going back to New York!" Terrace left that day, and Nim never saw him again. After Terrace's departure, Nim lay still in his cage, refusing food.

Raspy and long-haired—he wouldn't look out of place in a Santa Cruz head shop—Bob Ingersoll was the last and most enduring force for good in Nim's life at a time when everyone else, including Terrace, seems to have given up on him. Ingersoll rallied to get Nim out of the medical research facility to which an insolvent Lemmon had sold him and many other chimps, and again came to Nim's aid after he wound up sequestered at the Black Beauty Ranch, a rescue home for horses in Texas run by an activist with good intentions and no clue about how to care for a chimp. There is footage of Nim alone in a concrete room, maniacally shoving a metal barrel around on the floor. These images are capable of testing the empathy of even the most rigid human exceptionalists. Don't call it anthropomorphism; the emotions this animal displays are unmistakable: sadness, bitterness, loneliness, betrayal, rage.

Marsh's documentary avoids the depths that lie beneath Terrace's bizarre, tragic experiment—things that could be said about linguistics, the imperfect nature of science, and some of the most interesting areas of philosophy. (Making a film about animal language without mentioning Noam Chomsky is a bit like writing a book on the French Revolution that neglects to bring up Louis XVI.) One of the narratives that remains largely untold is the devastating effect the experiment had on the future of research in the field.

Terrace made himself one of the most powerful enemies of such research when he declared, in a 1979 paper in *Science* entitled "Can an Ape Create a Sentence?" and in a related book, that Project Nim and by extension *all* animal-language experiments were bunkum—the wishful thinking of sloppy scientists deceived by their subjects' clever and complex ways of begging for treats. Ape-language research has yet to recover from Terrace's public surrender to Chomsky—a turnaround that felt especially treacherous considering the inexactitude of Terrace's own science.

Many of the problems of Project Nim arose from Terrace's faithful acceptance of Chomsky's syntax-based definition of language, and from a resulting methodology rooted in the familiar techniques of second-language instruction. The word itself, "instruction," is indicative of a wrong-headed way of thinking about the acquisition of a first language. Human infants do not really need to be "instructed" in their first language to pick it up—this was at the heart of Chomsky's argument in the first place. When Terrace decided that the environment surrounding Nim was too chaotic, he resolved to have Nim instructed in a "classroom" at Columbia. The classroom was small, windowless, and whitewashed, with nothing in it except Nim, the person working with him, and some drab, minimal furnishings. The idea was to hone Nim's concentration by isolation. We see his caretakers struggling to maintain his restless attention. Joyce Butler, one of his keepers, tells of her realization that Nim was making the "dirty" sign—indicating that he had to use the bathroom—simply in order to get out of there. Such an environment makes little sense for what Terrace and his staff were trying to do: help a conscious being acquire a first language. Decontextualizing language from the everyday in order to foster its acquisition is like putting a seed in a sealed jar to help it sprout.

The case of Kanzi is a helpful counterexample. Beginning in the mid-1970s, Duane Rumbaugh and Sue Savage-Rumbaugh, who had also worked with Bill Lemmon at the University of Oklahoma, began a series of language experiments with chimpanzees and, later, bonobos—a cousin species about which relatively little was known at the outset of their research—employing a table of invented lexigrams: arbitrary, nonrepresentative pictures signifying certain things (actions, foods, places, the names of apes and people involved in the experiment, and so on). One reason for their creation of the lexigram system was to help alleviate the data-gathering problems of the sign-language experiments. Whereas both the Washoe and Nim experimenters struggled with ASL and its interpretation, Rumbaugh and Savage-Rumbaugh had only to decide whether an ape was touching a picture that was neatly blocked off from others in a little square, which is a much more objectively measurable datum. In the early 1980s, the couple were trying to get Matata, an adult female bonobo, to understand and use the lexigrams. Matata had recently stolen an infant named Kanzi from a bonobo in captivity. Kanzi was either close by or clinging to his adoptive mother while the research was going on. They never had much luck getting Matata to understand the lexigrams, but later realized that her son had picked up many of their meanings—spontaneously, and with no deliberate instruction. That is, in the same way one acquires a first language.

Since then, the experiment has expanded along with the group of bonobos involved in it. It is the only ape-language experiment still active, and is currently based at the Great Ape Trust's research facility in Des Moines. Many of the bonobos understand not only the lexigrams but also a great deal of spoken English. One can view online videos of Kanzi carrying out simple tasks at spoken request. The experimenter—wearing a welding mask to help prevent unintentional facial cueing—might say, for instance, "Put the snake on the ball," and Kanzi responds by placing a toy snake on top of a plastic beach ball, suggesting comprehension of verb, word order, and preposition. The room is strewn with these various artifacts —balls, soap, stuffed animals, water pails, plants—resembling far less Nim's stark isolation chamber than, say, a nursery: a space including not only words, but things to talk about.

Researchers who conduct language experiments with animals—especially complex, social, intelligent ones like great apes—sometimes draw the public's interest, but after Project Nim they have had a hard time persuading the scientific community to consider their work anything more than wishful thinking. Few have done more to aggravate animal language's respectability problem than Herb Terrace himself. For ape language's skeptics, he provided the voice of the disgruntled inside man. In 1980, shortly after Terrace published his paper denouncing the project in *Science,* the Indiana University linguist Thomas Sebeok organized a conference on the "Clever Hans phenomenon." The term refers to humans anthropomorphizing animal behavior in such a way as to assume cognitive or communicative processes that aren't really occurring. (Clever Hans was a horse who, his trainer believed, could answer simple mathematical equations, among other things, by stamping out solutions with his hoof; by isolating the horse from his owner, researchers found that Clever Hans was determining his "answers" by picking up subtle subconscious cues from his human—still impressive, to be sure, but the horse was not doing arithmetic.) Sebeok invited Herbert Terrace to the Clever Hans conference, during which he stridently pronounced that "the alleged language experiments with apes divide into three groups: one, outright fraud; two, self-deception; three, those conducted by Terrace."

"The combined effect of Sebeok's Clever Hans Conference and Terrace's *Science* paper," Sue Savage-Rumbaugh later wrote,

> was . . . to instigate an extremely rapid and violent swing of the pendulum. Ape-language research went from being a field of perceived intellectual excitement and public acclaim to one that, at best, should be viewed askance. Suddenly, it became extremely difficult to have research papers reviewed, let alone published. And funding for most of the major projects virtually dried up.

The end of Project Nim marked the end of an incautious but intensely curious open-mindedness in the culture of science that was probably reflective of a change in culture at large: free-spiritedness was out, and the skeptical, cynical eighties were in. Jenny Lee's remark ("It was the seventies") not only calls attention to the look of the film—all the grainy and garishly colorful footage of chimp caretakers, knee socks, bell-bottoms, and sideburns—but also suggests that this story could not have happened in the way it did at any other time. The backdrop of hedonistic abandon behind this story may strike a viewer today as humorous or appalling. One wonders how much serious scientific inquiry was going on in an environment in which everyone was in bed with everyone, and Nim was plied with booze and pot right from infancy. To watch a chimpanzee puffing on a joint is disquieting, in equal measures funny and disturbing. We enjoy mocking that sliver of biological difference between us and chimpanzees. Yet anyone who has ever looked with curiosity and respect into the face of a chimpanzee has seen a presence there. If we abandon the notion that language is necessarily the bedfellow of consciousness, we get a better understanding of ourselves, while our relationship to the other beings we share this planet with becomes more enlightened, more humble, and more humane.

TIM ZIMMERMANN

Talk to Me

FROM *Outside*

STRETCHING NORTH AND east from Grand Bahama Island, the Little Bahama Bank is a vast, crescent-shaped undersea plateau of sugar-white sand, patchy seagrass, and isolated coral reefs, layered under a shallow veneer of translucent water. It sits just 60 miles east of West Palm Beach, across the Gulf Stream. Yet despite its proximity to the condo sprawl of Florida, it is another world, a wild seascape of endlessly changing water and light, fast-moving thunderstorms, and teeming bird and sea life.

My first contact with its alien underwater culture involved a snorkel, a mask, and fins. I dropped into the 83-degree sea, and on the periphery of my vision six sleek shapes wheeled and turned, gliding with perfect ease. Three were larger and mottled with spots. The others were colored a smooth, gunmetal gray. One broke formation and arrowed my way, scanning me with a sophisticated sensor system. I heard a high-pitched buzz that sounded like a zipper being ripped open and could feel a light vibration in my chest. As the creature shot past, it rolled slightly to make direct and steady eye contact.

The scientific name for the species is *Stenella frontalis*. The more common name is the Atlantic spotted dolphin. There is a group of about a hundred of them living near the western edge of the Little Bahama Bank, and for the past twenty-eight years Denise Herzing, a marine-mammal biologist in her midfifties, has devoted her life to learning about them and their culture. Since 1985, she has spent close to one hundred days every summer here, enduring baking sun and nosy sharks so she can observe their wild society.

At this point, she recognizes about sixty of the dolphins by sight. (The others she identifies using her photo catalog.)

Over the years, Herzing has had close to 2,500 encounters with these dolphins and spent some 1,500 hours in the water with them, accumulating research for the Wild Dolphin Project, a nonprofit in Jupiter, Florida, that she founded in 1985. She has an extensive video and sound library of the clicks and whistles the dolphins use to communicate. She has also learned intimate details about their complex world—how males form tight coalitions and cruise the waters like scrappy gangs; how young females babysit calves to prepare for motherhood; how everyone seems to have sex (or at least play at sex) with everyone. "It's really interesting to see what's going on in the mind of another species," says Herzing, who is an affiliate assistant professor in the biological sciences department at Florida Atlantic University and has written or collaborated on some thirty scientific papers about the dolphins. "They have the potential to show you their world in real time."

Now Herzing plans to take her relationship with the spotted dolphins to an ambitious new level. She is refining a set of portable underwater communication devices that can recognize and generate dolphinlike whistles, and she plans to use them to establish two-way communication. She'll start by exposing the dolphins to a few of the whistles, using pattern-recognition software to tell her, via earphones she'll wear underwater, if they use them to whistle back. Herzing hopes that once the dolphins, who are skilled mimics, get the idea, they can build a communication system together. "Maybe it will lead to an extensive artificial language," Herzing says. "But the real breakthrough would be if the dolphins introduce their own vocalizations and whistles."

It's a radical goal. Herzing the scientist is trying to achieve something that has never been done before: two-way communication with a wild species. Herzing the person has a more existential aim: to open up an entirely different view of the planet and its creatures that is not so monumentally human-centric. "I think it could be our salvation," she says. "Because if we don't start including other creatures in the formula, there is not going to be a planet."

To launch this grand experiment, we will spend ten days at the Little Bahama Bank on the *Stenella*, the 62-foot power catamaran Herzing uses as her oceangoing research base. Herzing has two experienced dolphin researchers, two graduate students, and two computer techs on board to help her. A captain, a mate, and a cook keep the boat running smoothly so she can focus on her work. Inside, there's a well-stocked galley, twelve bunks, a large lounge area to review video footage and log dolphin data, and generators to power the electronic gear and the air conditioner.

We've dropped anchor in 14 feet of water. No dry land is visible. We are alone—save for the occasional passing boat—in a vast ocean wilderness, rocked gently by a building swell.

Humans have always been fascinated by the idea of communicating with other species, and the past forty years have seen some impressive breakthroughs. Koko, a gorilla born in 1971 at the San Francisco Zoo, learned American Sign Language and knows more than 1,000 signs. Kanzi, a bonobo chimpanzee at the Great Ape Trust in Iowa, mastered a keyboard with more than 300 lexigrams and can understand some 3,000 words of spoken English. Alex, an African grey parrot that lived from 1976 to 2007, could vocalize about 100 English words and count to six.

Harder to grasp is the idea that dolphins, nonprimates that live in water and have been on a separate evolutionary track from humans for 95 million years, might be capable of two-way communication. Humans and their more evolved primate cousins, it was long assumed, were unique in possessing the necessary intelligence for sophisticated communication. But the two other dolphin cognition experts aboard the *Stenella*—Adam Pack, forty-nine, an associate professor at the University of Hawaii, and Matthias Hoffmann-Kuhnt, forty-seven, an acoustics specialist living in Singapore—know more than most about how wrong that assumption was. They met at the University of Hawaii's Kewalo Basin Marine Mammal Laboratory, founded by Louis Herman, a psychology professor and former air force intelligence officer. From the 1970s through the 1990s, Herman masterminded a series of experiments with his bottlenose dolphins, which lived in two interconnected saltwater pools near the beach in Honolulu and demonstrated startling language and cognitive skills. Herman's dolphins learned the meaning of more than thirty signals, including nouns and verbs. They also mastered syntax and grammar rules. If a trainer signaled "person, surfboard, fetch," the dolphins would bring the surfboard in the tank over to the person.

But if the trainer changed the order of the words and signaled "surfboard, person, fetch," the dolphins knew to bring the person to the surfboard.

The Kewalo dolphins were also able to grasp abstractions. They understood the difference between left and right, could comprehend the existence of an object even if it wasn't present, and correctly responded to a trainer shown on a television screen, understanding that it was a representation of the real world.

Then, a decade ago, another extraordinary fact of dolphin intelligence was established. For centuries, humanity basked in the egocentric belief that humans alone were self-aware. In the 1970s, however, Gordon Gallup Jr., a researcher at Tulane University in New Orleans, used a mirror to show that chimpanzees are also self-aware and can recognize that the image they see is not another animal but themselves. In the 1990s, Diana Reiss, now a professor of psychology at New York City's Hunter College, in collaboration with Lori Marino, a doctoral student in Gallup's lab, started using the mirror test on dolphins. By the way the animals acted, they, too, demonstrated that they were seeing themselves. To corroborate, Reiss used a nontoxic black marker to mark the dolphins: when one swam past the mirror for the first time with a mark on its head, Reiss says, it did a "classic double take" and immediately returned for a look. A human child starts to react to a mirror in the same way around age two.

Published in 2001, it was a breakthrough study, and dolphins became the first nonprimate species to show evidence of self-awareness. The work affected Marino deeply. "Despite being so different in terms of how their brains are organized, and where they live and what they look like, dolphins show a surprising degree of similarity to humans in terms of the kind of self-awareness they have," says Marino, now a senior lecturer in the neuroscience and behavioral biology program at Atlanta's Emory University.

Marino, who has worked with Herzing, concluded that it wasn't ethical to keep and study such an intelligent creature in captivity and switched to using MRI and other imaging technology to analyze the size and structure of brains from dolphins that have died of natural causes. One of the simplest ways to get a sense of brain potential is the encephalization quotient, which compares real brain mass with the brain mass expected for a given body size. An EQ of 1, for example, means the brain mass is what you would

expect for the body housing it. Humans have the highest measured EQ on the planet, at 7, meaning our brains are seven times
larger than our body size would predict. Chimpanzees, our closest relatives, come in with an EQ of 2.3. Marino's calculations on
dolphins showed that bottlenose dolphins have an EQ of 4.2, and
dolphins on average have a higher EQ than primates. In short,
Marino showed that dolphins are second only to humans when it
comes to brain complexity.

Herzing, who has worked with dolphin researchers from all
over the world, is respectful of captive research. But diving into
the dolphins' world is her preferred way to learn about them.
(Most captive research has involved bottlenose dolphins; Herzing
works with spotteds because they happen to be the dolphins she
has the fullest access to in the wild.) "A captive dolphin can tell
you things about its cognition," says Adam Pack, who also studies
humpback whales. "But a wild dolphin can teach you things about
its culture. And to have a dolphin assist you in understanding what
other dolphins are doing is an area that we haven't ever gotten
into."

Herzing has named each spotted dolphin at the Little Bahama
Bank, using monikers like Little Gash, Linus, and Pointless (for a
dolphin that lost its dorsal tip, likely to a shark). When they race
in to ride the bow wave of the *Stenella,* she enthusiastically greets
them with a wave and a whistle, grinning as they roll onto their
sides to look up at her. One afternoon we watch underwater as
Deni, a young female, tries to teach Cobalt, a new calf she is babysitting, how to fish. Deni uses her beak to show Cobalt how to dig a
garden eel out of the sand, and then she deftly helps keep the eel
in front of Cobalt as he learns to track its frantic course. Suddenly,
Bijyo, a juvenile female, glides in and gulps down the eel, bringing the lesson to an abrupt and unscheduled end. Deni appears
outraged, as if Bijyo has committed a serious offense against the
sanctity of babysitting. She goes after Bijyo, opening her mouth in
an aggressive show, and the two start twisting and turning in the
water.

Herzing, Pack, and others believe that such structured social
patterns may help explain the dolphin brain. "It's sometimes
called the Machiavelli hypothesis, and it is that individuals who live
in complex social groups require complex cognition," says Stan
Kuczaj, a professor of psychology at the University of Southern

Mississippi, who has been out on the *Stenella* with Herzing. "To thrive, you have to understand what the social rules are—when you have to obey them and when you can get away with not obeying them—and who the players are."

The desire to better understand the dolphin brain led to efforts to communicate with the animals. Starting in the 1980s, Reiss and a few others, including Kuczaj, introduced keyboards and artificial sounds into their research. (In the late 1990s, Herzing, with Pack's help, also attempted to use a whistling keyboard at the Little Bahama Bank, but it proved too clunky.) The captive research showed that dolphins could mimic and understand artificial whistles and would perhaps even use them on their own.

Herzing wants to build on that work by exposing wild dolphins to artificial whistles and associating those whistles with specific toys, in the hope that the animals start using the whistles to request the toys. The holy grail would be if, over the years, the whistle vocabulary developed to a point where the "conversation" might include social needs and insights into the dolphin world, with the spotteds perhaps communicating about things like predators, family, or sex.

"Why Denise's work is exciting is that none of the dolphin studies have been truly two-way," says Heidi Lyn, also a professor at the University of Southern Mississippi, who has worked with both Kanzi, the bonobo chimp, and the Kewalo dolphins. "The humans could request or the dolphins could request, but Denise is working to achieve a back-and-forth. And if the dolphins introduce their own whistles, that would be amazing."

On our first afternoon at the Little Bahama Bank, the *Stenella* is swinging to her anchor under a humid sky. Herzing, wearing a bathing suit, visor, and dark glasses, is on the back deck, preparing to test the communication equipment in the water. Everyone has to be ready to don snorkeling gear and jump in any time a dolphin appears. (Herzing doesn't scuba-dive, because the bubbles distract the dolphins and the apparatus is too bulky for a fast gear-up.) "It's kind of like being a fireman," Herzing told the group during a briefing the first night. "You sit around a lot, and then there's a sudden rush."

To build the underwater communication devices, Herzing partnered with the Wearable Computing Lab at Georgia Tech, which

I visited with her last April. There's no shortage of futurism there: when you step out of the elevator, two eyeballs stare at you from a screen and track you as you walk away. The lab is run by Thad Starner, a PhD in computer science from MIT who, together with a former student, developed a superfast pattern-recognition algorithm. "To really prove the value of the algorithm, we hope to use it to help Denise learn something new about dolphin vocalization," Starner says.

Starner and his team have dubbed the communication devices CHAT boxes, for Cetacean Hearing and Telemetry. Two of Starner's whiz kids—Stewart Butler, a beefy twenty-three-year-old computer science major, and Daniel Kohlsdorf, a twenty-six-year-old PhD student, both of whom are on the *Stenella*—have spearheaded work on a prototype for months. Each unit consists of a milled aluminum box, about the size of a laptop, that contains a cell-phone processor and is loaded with pattern-recognition software written by Kohlsdorf. Wired to the box are an underwater speaker, two hydrophones, and a keypad. Buttons on the keypad allow a user to emit artificial whistles at frequencies within both human-hearing and dolphin-vocalization range—which extends at least ten times beyond a human's. For the initial work, Herzing has matched specific whistles to toys she knows the dolphins like: a rope, a scarf, and sargassum, a common local seaweed that dolphins often play with.

The whistles should be easy for the dolphins to mimic, though hoping they'll confine their responses to the limited human range is sort of like asking them to speak very low and slow, the way a person might when trying to communicate with a foreigner. The hydrophones serve as underwater "ears," and the pattern-recognition software has been programmed to identify the defined whistles—and to allow for some variation in frequency and modulation for a potential dolphin "accent." Whenever the box detects a designated whistle amid all the other chatter that often accompanies dolphin encounters, it will convert it into the assigned English word and pipe it into the snorkeler's ear via the underwater earpiece. Starner's team is also working to embed an LED system into a specialized mask, programming it to indicate where a whistle is coming from, to give Herzing a better sense of which dolphin is vocalizing. "The real-time communication is key," says Herzing, a dynamic and direct woman who loves to laugh and has short, sun-bleached hair. "Imagine trying to have a conversation where you go away to figure out what was said and come back twenty-four hours later and try to pick it up again."

Herzing is joined on the aft deck by Pack. Butler and Kohlsdorf bring out the equipment, which is far from streamlined at this stage, giving the whole exercise a garage-invention feel. Herzing and Pack don yellow vests that have Velcro straps to snug the CHAT boxes against their chests. Once the boxes are secure, the keypads are strapped to their forearms. "First ve deploy zee Denise, zen ve deploy zee Adam," declares Hoffmann-Kuhnt in an exaggerated German accent. This is his first time out on the Little Bahama Bank. His lab in Singapore builds all sorts of underwater gear, and he's an improvisational whiz. Later he'll layer cut-up pieces of a Mini-Wheats cereal box to the interior of the CHAT devices to absorb any leaks.

Herzing's strategy is to engage the dolphins by having two swimmers in the water play a game with one of the toys while using the CHAT boxes to whistle the sound for that toy. Dolphins, like children, are very good at learning by observing and then joining in. "If they really get it, I think they'll go down to the bottom and grab something like a sea cucumber and bring it to us with a whistle," Herzing says. "The intention is to convey that we want to interact and that we have the tools."

Herzing and Pack press the buttons on their keypads, and a series of distinctive trills cuts through the still air. Then they put on their masks and snorkels and stagger to the swim platform as the *Stenella* rolls. "The sea was angry that day, my friends, like an old man trying to send back soup in a deli," Pack jokes, channeling *Seinfeld*. He and Herzing trundle off the swim platform at the back of the boat, splashing into the water. Within seconds, both pull their heads out and shake them. As soon as the boxes were immersed in the sea, they quit. "That's why they put the 're' in 'research,'" Pack quips as he clambers back aboard.

Butler and Kohlsdorf gather up the boxes and take them into the makeshift workshop they've set up in the *Stenella*'s main lounge, which is quickly being overrun with spare parts and tools. The battle of Georgia Tech versus the ocean, otherwise known as prototype development, commences. Herzing is patient. "It's counterintuitive to put a computer in salt water," she says.

*

Three days later, after a night in West End to escape rough weather and conduct a troubleshooting conference call with Georgia Tech, we're back on the bank. Suddenly, Captain Pete Roberts shouts "Dolphins!" from the bridge. He toggles an alarm and stomps on the floor to alert everyone on the boat. Dark shapes are moving through the water to take up station in the pressure wave pushed up by the bow. It's a free ride, and the dolphins love hanging there, adjusting their position with almost invisible movements of flukes and fins. Jessica Cusick and Bethany Augliere, graduate students at Florida Atlantic, who have been working with Herzing on the Wild Dolphin Project for three years, immediately start calling out names and taking photos. (Herzing updates her ID catalog every year as the dolphins grow and their spot patterns change.) Once they have what they need, Herzing decides whether the conditions are right for a jump.

Butler and Kohlsdorf are still battling the boxes, pulling all-nighters and working their way through a fifth of Eagle Rare bourbon. They are being utterly confounded by small leaks, the capacitive oddities of salt water, and the gremlins that pop up when you field-test software for the first time. They adjust the code, tweak the hardware, the ground wiring, and the waterproofing, reassemble the boxes, and then dunk them into a large plastic vat of fresh water on the back deck to rinse them. Sometimes the boxes work, but usually not for long.

"What the fuck?" becomes the most common technical question on the *Stenella*. Kohlsdorf, who was adopted from Korea by German parents and recruited by Starner from the University of Bremen for his coding skills, is also smoking his way through most of a pack of Camels each day. He has long, shaggy black hair and a sense of fatalism that serves him well. "That's how it goes," he likes to say. But his T-shirt subtly contradicts his poise. It's black, and across the chest is a phrase in German. The translation: "It's also shitty somewhere else."

Herzing calls for a jump. She and Pack will take a scarf to practice the game that they'll model for the dolphins once the CHAT boxes are operational. Herzing's two-way strategy relies on the animals' love of play. She has to get them thinking, Hey, I want to get in on that game, and they have to understand that the way to do it is to mimic the whistle for "scarf" or whatever the play object is.

"As with kids, there is a lot of stuff that goes on before you achieve recognition," Pack explains. "They have to see that the system is fun, that it is functional, and that they can use it to ask for things. So it's a process."

We all slide into the water. A small group of spotteds is milling around nearby. Herzing swims out, a red scarf visible in her hand, with Pack trailing. I can see the dolphins take note of her. She drops the scarf and points at it. Bijyo, the garden-eel poacher, swims by and plucks the scarf out of the water column, with two others trailing her. She drops the scarf in front of Pack, who points at it, and Bijyo grabs it again. Bijyo seems to enjoy being the center of attention. The scarf drops from her mouth, and I think she's lost it—but then it catches on her pectoral fin, fluttering there as she cruises around.

Lucky, I think. But then I see the scarf slide off her pec, only to be picked up by her tail fluke. Luck has nothing to do with it, I realize. Bijyo, with impressively casual dexterity and awareness, has passed the scarf down her body. She finally drops it on the white sand. Then another spotted swoops in and with equal precision picks it up, using a tail fluke. Other dolphins show up, until there are more than a dozen swirling around. The scarf game stops only when one of them eventually swims off with it.

It's easy to see how the CHAT boxes will add an intriguing dimension to the proceedings. "Most social beings learn about each other through interaction," Herzing says later, when we review the session on video. Pack adds, "You model the behavior for the player, but the others are watching. It's like a classroom, but there is so much else going on, you can see how challenging it is. In the marine pool, you know they will be there at eight in the morning. Out here you have to hope you find them and, if you do, that the same players will turn up."

In the immensity and isolation of the Little Bahama Bank, you have to pause to figure out what day it is. Most of the time, the *Stenella* zigzags across the water, checking in on all of Herzing and Captain Pete's favorite spots. Roberts has been working with Herzing for twelve years, and he knows this area like his own backyard. It's never long before he finds dolphins and we're in the water. When the sun goes down, margaritas and beers come out and the grill gets fired up. Any new dolphin data or interactions are logged in exquisite detail, and video is reviewed. When there is time to relax, the talk often turns to the science of dolphin cognition.

The idea that dolphins are the humans of the sea (which is what the Maori called them), and that there is a special connection between humans and dolphins, has existed for centuries. Herzing herself once watched Jumper, a female, break off what she was doing to escort an exhausted swimmer back to the boat. Another time, the dolphins were behaving strangely and would not approach the *Stenella*. The crew on board soon discovered that one of the passengers had died quietly in his cabin, from a heart attack. As the *Stenella* motored toward West End, the dolphins swam for a way in escort, about 100 yards out on either side.

In the 1960s, John C. Lilly, a freethinking neuroscientist and friend of Timothy Leary and Allen Ginsberg, took the idea of a unique and potentially transformative bond between humans and dolphins to an extreme. Lilly, captivated by the intelligence and gentle nature of dolphins, believed that one of them could learn to speak English. He bought a house on St. Thomas, in the Virgin Islands, partially flooded it with seawater, and got his assistant, Margaret Howe, to live in the house with a bottlenose called Peter. Peter was incapable of producing the consonants and other sounds needed for English, but he did display an ability to closely mimic the patterns of Howe's speech. The experiment, unsurprisingly, deteriorated along with the hygienic conditions and Howe's tolerance for living alone with a dolphin. Lilly's extreme methods — he also tried giving dolphins LSD — have colored efforts to communicate with the animals ever since.

"He was a visionary, ahead of his time," Herzing says of Lilly as we cool off one afternoon in the air-conditioned lounge. "But he really lost the scientific process and decided to go off and explore his own mind with drugs. It has held two-way work with dolphins back for two decades, because people have been scared to death to be called another Lilly."

Nonetheless, the idea of a deep connection with dolphins has motivated Herzing's work from the start. She grew up in landlocked Minnesota and developed an obsession with the ocean by watching Jacques Cousteau and reading the *Encyclopaedia Britannica*. Even as a young girl, she knew she wanted to become a marine biologist and study dolphin communication. If she could do one thing for the world, she declared in an essay at age twelve, it would be to "develop a human-animal translator so we could understand other minds on the planet."

In the early 1980s, after earning a marine-biology degree from Oregon State University and traveling the world, Herzing realized she wanted to study dolphins in their natural environment instead of in captivity. She reasoned that if the best way to understand humans was in the context of human society, networks, and relationships, the same was likely true of dolphins. She wanted to study dolphin cognition and communication as an anthropologist might—in a natural setting. "I wasn't as interested in doing experiments," she recalls. "I was interested in observing, which is the most productive if you want to understand their culture."

It wasn't long before she became aware of a group of friendly spotted dolphins in the Bahamas, first noted by treasure divers in the 1970s. The water was warm and shallow, and the dolphins appeared to be easily accessible, even curious about humans. "I couldn't believe no one was studying them already," Herzing says. She began her fieldwork in 1985, which grew into the Wild Dolphin Project. Its motto captures Herzing's research ethos perfectly: "In their world. On their terms."

Still, her plan to develop two-way communication with the spotted dolphins presents an important ethical dilemma. Diana Reiss, for one, wonders whether there's a danger of somehow changing or harming a wild culture by bridging into it. "You have the potential to learn something you would never be able to learn, because they have the potential to show you something in their world that they wouldn't in an aquarium," she says. "But the question is, should we introduce new vocal elements into a wild population? And if we do, are we somehow contaminating their vocal repertoire? It's a basic philosophical question."

Herzing worries about that quite a bit. "The potential is that they will start showing you their world in detail, and you can have some interface that will help you understand the wild mind," she tells me one morning after sunrise as we sit on the *Stenella*'s bridge. "The danger is that you could get too much into their system. That you could have young animals who spend too much time with humans and don't do the things they are supposed to do, putting pressure on the mothers. Or you could get dolphins who get to trust humans and a bad human comes along."

In a way, it's like the search for extraterrestrial intelligence, except here on Earth. The spotted dolphins are like aliens that inhabit a completely different world. Scientists at the SETI Institute, in Mountain View, California, are, in fact, paying close attention to Herzing's work. "What we're trying to do with SETI is communicate with a different form of intelligence," says Doug Vakoch, the institute's director of interstellar message composition. "Denise's work has highlighted some of the things we need to take into account, like the importance of interactivity and a long-duration relationship."

Inevitably, these attempts are open to ridicule. Rush Limbaugh, who somehow got wind of Herzing's project, joked to listeners in May: "The dolphins can tell us how they keep health care costs down without Obamacare and how they avoid trillion-dollar deficits." Herzing's scientific colleagues respect her dedication and rigor, and when I ask Herzing if people think she's a kook, she's not offended. "No," she answers. Then she adds, "At least not yet," and cuts loose with one of her signature cackles.

Dusk is falling on Monday evening, our sixth day out. Butler and Kohlsdorf emerge onto the back deck carrying two complete CHAT boxes. They have made progress, but the boxes have to be totally reliable before Herzing will introduce them to the dolphins. "Having them quit in the middle of a session would be very confusing," she says. There is a faint air of hope, but after repeated false starts everyone has learned to live by Kohlsdorf's favorite response to all questions about whether the latest fix will work: "We will see."

Herzing and Pack gear up, and the water is dark enough that Captain Pete drops a light off the stern. Herzing and Pack slip into the water for what seems like the fiftieth test, ten feet apart in the green glow of the droplight. Herzing's box seems to work fine. After a few seconds, Pack lifts his head out of the water and says, "My unit is not playing anything."

Back on deck, Herzing is philosophical. "Shit happens," she says. "It's a tough environment out here." The next morning, the carcass of Pack's CHAT box lies open in the lounge. Water found a way in and drowned the components. Throughout the day, Herzing, Hoffmann-Kuhnt, and Kohlsdorf discuss the modifications they'll make to improve waterproofing, software stability, and electrical grounding. Kohlsdorf is wearing a different black T-shirt. It depicts a toddler in a Jason-style hockey mask, dragging a bloody chain saw.

If and when the CHAT system is finally debugged, it's likely the dolphins will be ready to show Herzing something new. The next morning, we drop in on a curious scene featuring three bottle-nose dolphins and a group of seven spotteds. About a hundred bottlenoses share the Little Bahama Bank with the spotteds, but they are more skittish, and Herzing hasn't spent as much time with them. After days of looking at friendly little spotteds in the water, the bottlenoses appear enormous and a bit menacing, like outlaw bikers. With penises visibly erect, they have been mauling a young spotted male named Lhasa, while Lhasa's three buddies—Linus, Malibu, and Kai—hover nearby. When I get into the water, two bottlenoses swim up toward me, their pale erections still waving in the water. Don't get confused, boys, I think nervously.

Instead, they turn away and direct their attention to Malibu. They swarm around him, trying to jam themselves into him. Malibu twists and turns, but he doesn't try to flee. The other spotteds follow the action but don't intervene. Eventually, the bottlenose dolphins break away and swim off. At that point, four of the male spotteds abruptly turn and, in tight formation, swim right up to me, so we are all eye to eye. They are like four authoritative bouncers. With my puny human hearing I can't know if they have anything to say to me. But their posture and eyes alone convey a simple and direct message: Enough already, human voyeur. It's time for you to leave. With that, they spin away and disappear. I head for the swim ladder.

"That's some crazy shit," I say on deck, abandoning any pretense of scientific inquiry. "What's up with that?" Herzing agrees that it is pretty remarkable behavior, but she and Pack don't really have an explanation. Over the years, Herzing has seen lots of bottlenoses hanging around with spotteds, and vice versa, and she has also seen evidence of limited interbreeding. Despite the controlled aggression, what struck me was the ritualized nature of the interaction. The spotteds never tried to fight or flee; most just hung around as if observing a frat-house hazing. It is clearly an interesting relationship. There is confrontation, but it has limits, and Herzing doesn't know for sure, but she has never seen aggression between spotted and bottlenose dolphins escalate to killing. Pack is also intrigued by the sympatric relationship that the spotteds and bottlenoses have, sharing the same territory yet coexisting and maintaining their distinct cultural and genetic identities without extreme violence. It's a thought-provoking, perhaps even instructive, model.

Before we head back to Florida, we make one last drop with a large group of spotteds. I count twenty-four, traveling slowly across the sand flats, a community on the move. The ever-excitable calves sometimes dart away, only to be chased down by their mothers or babysitters and firmly set back in the pack. Groups of males are swimming in close formation, keeping an eye on us. Deni and Bijyo are still hassling one another. Small groups break off to chase fish out of the sand. The entire community forms and re-forms in a hundred subtle ways, and I can see lots of pec touching, the dolphin equivalent of a reassuring hand. I know I can't comprehend 99 percent of the social dynamics and communication in play, but it's also impossible not to feel that they're there and worth trying to fully understand.

Herzing will spend another seventy days on the Little Bahama Bank this year, and more in the years to come, refining the CHAT system. Stan Kuczaj and Diana Reiss, among many others, will be watching, to see how far she can take it. "There is a difference between communicating and engaging in a meaningful conversation," Kuczaj says. "Conversations require shared interests, and finding common ground may be more difficult than we imagine."

If any human can find that common ground, it's probably Herzing. She is arguably more connected with a wild dolphin culture than anyone else on the planet. One evening I ask Pack if he's confident that a few decades from now we'll have cracked the code of dolphin communication. He thinks and then answers: "As long as we're on a positive trajectory and have technology, we'll understand more of what the code is. It may not be what we thought it would be, but we'll have a general understanding."

The boat settles in for the night. Overhead there is a thick carpet of stars, and a warm wind whispers over the undulating water. I know the spotted dolphins are somewhere nearby, perhaps headed into deep water to feed on squid. Their world seems both separate and connected to ours, and it's suddenly easy to believe that something extraordinary will happen if Herzing builds a meaningful two-way connection between humanity and the wild and alien culture that thrives on the Little Bahama Bank.

DAVID DEUTSCH AND ARTUR EKERT

Beyond the Quantum Horizon
FROM *Scientific American*

LATE IN THE nineteenth century, an unknown artist depicted a traveler who reaches the horizon, where the sky meets the ground. Kneeling in a stylized terrestrial landscape, he pokes his head through the firmament to experience the unknown. The image, known as the Flammarion engraving, illustrates the human quest for knowledge. Two possible interpretations of the visual metaphor correspond to two sharply different conceptions of knowledge.

Either it depicts an *imaginary* barrier that, in reality, science can always pass through, or it shows a *real* barrier that we can penetrate only in our imagination. By the latter reading, the artist is saying that we are imprisoned inside a finite bubble of familiar objects and events. We may expect to understand the world of direct experience, but the infinity outside is inaccessible to exploration and to explanation. Does science continually transcend the familiar and reveal new horizons, or does it show us that our prison is inescapable—teaching us a lesson in bounded knowledge and unbounded humility?

Quantum theory is often given as the ultimate argument for the latter vision. Early on, its theorists developed a tradition of gravely teaching willful irrationality to students: "If you think you understand quantum theory, then you don't." "You're not allowed to ask that question." "The theory is inscrutable and so, therefore, is the world." "Things happen without reason or explanation." So textbooks and popular accounts have typically said.

Yet the developments of the past couple of decades contradict those characterizations. Throughout the history of the field, physicists often assumed that various kinds of constraints from quantum physics would prevent us from fully harnessing nature in the way that classical mechanics had accustomed us to. None of these impediments have ever materialized. On the contrary, quantum mechanics has been liberating. Fundamentally quantum-mechanical attributes of objects, such as superposition, entanglement, discreteness, and randomness, have proved not to be limitations but resources. Using them, inventors have fashioned all kinds of miraculous devices, such as lasers and microchips.

These were just the beginning. We will increasingly use quantum phenomena for communications and computation systems that are unfathomably powerful from a classical point of view. We are discovering novel ways of harnessing nature and even of creating knowledge.

Beyond Uncertainty

In 1965 Intel's cofounder Gordon Moore predicted that engineers would double the number of transistors on a chip every two years or so. Now known as Moore's Law, this prediction has held true for more than half a century. Yet from the outset, it rang warning bells. If the law continued to hold, you could predict when transistors would reach the size of individual atoms—and then what? Engineers would enter the realm of the unknowable.

In the traditional conception of quantum theory, the uncertainty principle sets a limit that no technological progress could ever overcome: the more we know about some properties, such as a particle's position, the less we can know about others, such as the particle's speed. What cannot be known cannot be controlled. Attempts to manipulate tiny objects meet with rampant randomness, classically impossible correlations, and other breakdowns of cause and effect. An inescapable conclusion followed: the end of progress in information technology was nigh.

Today, however, physicists routinely exert control over the quantum world without any such barrier. We encode information in individual atoms or elementary particles and process it with exquisite precision, despite the uncertainty principle, often creating functionality that is not achievable in any other way. But how?

Let us take a closer look at a basic chunk of information, as traditionally conceived: the bit. To a physicist, a bit is a physical system that can be prepared in one of two different states, representing two logical values: no or yes, false or true, 0 or 1. In digital computers, the presence or absence of a charge on the plates of a capacitor can represent a bit. At the atomic level, one can use two states of an electron in an atom, with 0 represented by the lowest-energy (ground) state and 1 by some higher-energy state.

To manipulate this information, physicists shine pulses of light on the atom. A pulse with the right frequency, duration, and amplitude, known as a π-pulse, takes state 0 into state 1, and vice versa. Physicists can adjust the frequency to manipulate two interacting atoms, so that one atom controls what happens to the other. Thus we have all the ingredients for one- and two-bit logic gates, the building blocks of classical computers, without any impediment from the uncertainty principle.

To understand what makes this feat of miniaturization possible, we have to be clear about what the uncertainty principle does and does not say. At any instant, some of the properties of an atom or other system, called its observables, may be "sharp"—possess only one value at that instant. The uncertainty principle does not rule out sharp observables. It merely states that not all observables in a physical system can be sharp at the same time. In the atom example, the sharp observable is energy: in both the 0 and 1 states, the electron has a perfectly well-defined energy. Other observables, such as position and velocity, are not sharp; the electron is delocalized, and its velocity likewise takes a range of different values simultaneously. If we attempted to store information using position and velocity, we would indeed encounter a quantum limit. The answer is not to throw up our hands in despair but to make a judicious choice of observables to serve as computer bits.

This situation recalls the comedy routine in which a patient tells a doctor, "It hurts when I do this," to which the doctor replies, "Don't do that." If some particle properties are hard to make sharp, there is a simple way around that: Do not attempt to store information in those properties. Use some other properties instead.

Beyond Bits

If all we want to do is build a classical computer using atoms rather than transistors as building blocks, then sharp observables are all we need. But quantum mechanics offers much more. It allows us to make powerful use of nonsharp observables, too. The fact that observables can take on multiple values at the same time greatly enriches the possibilities.

For instance, energy is usually a sharp observable, but we can turn it into a non-sharp one. In addition to being in its ground state or its excited state, an electron in an atom can also be in a superposition—both states at once. The electron is still in a perfectly definite state, but instead of being either 0 or 1, it is 0 *and* 1.

Any physical object can do this, but an object in which such states can be reliably prepared, measured, and manipulated is called a quantum bit, or qubit. Pulses of light can make the energy of an electron change not only from one sharp value to another but from sharp to non-sharp, and vice versa. Whereas a π-pulse swaps states 0 and 1, a pulse of the same frequency but half the duration or amplitude, known as a $\pi/2$-pulse, sends the electron to a superposition of 0 and 1.

If we attempted to measure the energy of the electron in such a superposition, we would find it was either the energy of the ground state or the energy of the excited state with equal probability. In that case, we would encounter randomness, just as the naysayers assert. Once again, we can readily sidestep this apparent roadblock—and in doing so create radically new functionality. Instead of measuring the electron in this superposition, we leave it there. For instance, start with an electron in state 0, send in a $\pi/2$-pulse, then send in a second $\pi/2$-pulse. Now measure the electron. It will be in state 1 with a 100 percent probability. The observable is sharp once again.

To see the significance, consider the most basic logic gate in a computer, NOT. Its output is the negation of the input: 0 goes to 1, 1 to 0. Suppose you were given the following assignment: design the square root of NOT—that is, a logic gate that, acting twice in succession on an input, negates it. Using only classical equipment, you would find the assignment impossible. Yet a $\pi/2$-pulse implements this "impossible" logic gate. Two such pulses in succession have exactly the desired effect. Experimental physicists have built this and other classically impossible gates using qubits made of such things as photons, trapped ions, atoms, and nuclear spins. They are the building blocks of a quantum computer.

Beyond Classical Computation

To solve a particular problem, computers (classical or quantum) follow a precise set of instructions—an algorithm. Computer scientists quantify the efficiency of an algorithm according to how rapidly its running time increases when it is given ever-larger inputs to work on. For example, using the algorithm taught in elementary school, one can multiply two n-digit numbers in a time that grows like the number of digits squared, n^2. In contrast, the fastest known method for the reverse operation—factoring an n-digit integer into prime numbers—takes a time that grows exponentially, roughly as 2^n. That is considered inefficient.

By providing qualitatively new logic gates, quantum mechanics makes new algorithms possible. One of the most impressive examples is for factoring. A quantum algorithm discovered in 1994 by Peter Shor, then at Bell Laboratories, can factor n-digit numbers in a series of steps that grows only as n^2. For other problems, such as searching a long list, quantum computers offer less dramatic but nonetheless significant advantages. To be sure, not all quantum algorithms are so efficient; many are no faster than their classical counterparts.

Most likely, the first practical applications of general-purpose quantum computers will not be factorization but the simulation of other quantum systems—a task that takes an exponentially long time with classical computers. Quantum simulations may have a tremendous impact in fields such as the discovery of new drugs and the development of new materials.

Skeptics of the practicality of quantum computing cite the arduous problem of stringing together quantum logic gates. Apart from the technical difficulties of working at single-atom and single-photon scales, the main problem is that of preventing the surrounding environment from spoiling the computation. This process, called decoherence, is often presented as a fundamental limit to quantum computation. It is not. Quantum theory itself provides the means of correcting errors caused by decoherence. If the sources of error satisfy certain assumptions that can plausibly be met by ingenious designers—for instance, that the random errors occur independently on each of the qubits and that the logic gates are sufficiently accurate—then quantum computers can be made fault-tolerant. They can operate reliably for arbitrarily long durations.

Beyond Conventional Mathematical Knowledge

The story of the "impossible" logic gates illustrates a startling fact about the physics of computation. When we improve our knowledge about physical reality, we sometimes improve our knowledge of the abstract realms of logic and mathematics, too. Quantum mechanics will transform these realms as surely as it already has transformed physics and engineering.

The reason is that although mathematical *truths* are independent of physics, we acquire *knowledge* of them through physical processes, and which ones we can know depends on what the laws of physics are. A mathematical proof is a sequence of logical operations. So what is provable and not provable depends on what logical operations (such as NOT) the laws of physics allow us to implement. These operations must be so simple, physically, that we know, without further proof, what it means to perform them, and that judgment is rooted in our knowledge of the physical world. By expanding our repertoire of such elementary computations to include ones such as the square root of NOT, quantum physics will allow mathematicians to poke their heads through a barrier previously assumed to exist in the world of pure abstractions. They will be able to see, and to prove, truths there that would otherwise remain hidden forever.

For example, suppose the answer to some unsolved mathematical puzzle depends on knowing the factors of some particular enormous integer N—so enormous that even if all the matter in the universe were made into classical computers that then ran for the age of the universe, they would still not be able to factor it. A quantum computer could do so quickly. When mathematicians publish the solution, they will have to state the factors at the outset, as if pulled out of a magician's hat: "Here are two integers whose product is N." No amount of paper could ever suffice to detail how they had obtained those factors.

In this way, a quantum computer would supply the essential key that solves the mathematical puzzle. Without that key, which no classical process could realistically provide, the result would never be known. Some mathematicians already consider their subject an empirical science, obtaining its results not only by careful reasoning but also by experiments. Quantum physics takes that approach to a new level and makes it compulsory.

Beyond Bad Philosophy

If quantum mechanics allows new kinds of computation, why did physicists ever worry that the theory would limit scientific progress? The answer goes back to the formative days of the theory.

Erwin Schrödinger, who discovered quantum theory's defining equation, once warned a lecture audience that what he was about to say might be considered insane. He went on to explain that when his famous equation describes different histories of a particle, those are "not alternatives but all really happen simultaneously." Eminent scientists going off the rails are not unknown, but this 1933 Nobelist was merely making what should have been a modest claim: that the equation for which he had been awarded the prize was a true description of the facts. Schrödinger felt the need to be defensive not because he had interpreted his equation irrationally but precisely because he had not.

How could such an apparently innocuous claim ever have been considered outlandish? It was because the majority of physicists had succumbed to bad philosophy: philosophical doctrines that actively hindered the acquisition of other knowledge. Philosophy and fundamental physics are so closely connected—despite numerous claims to the contrary from both fields—that when the philosophical mainstream took a steep nosedive during the first decades of the twentieth century, it dragged parts of physics down with it.

The culprits were doctrines such as logical positivism ("If it's not verifiable by experiment, it's meaningless"), instrumentalism ("If the predictions work, why worry about what brings them about?"), and philosophical relativism ("Statements can't be objectively true or false, only legitimized or delegitimized by a particular culture"). The damage was done by what they had in common: denial of realism, the commonsense philosophical position that the physical world exists and that the methods of science can glean knowledge about it.

It was in that philosophical atmosphere that the physicist Niels

Bohr developed an influential interpretation of quantum theory that denied the possibility of speaking of phenomena as existing objectively. One was not permitted to ask what values physical variables had while not being observed (such as halfway through a quantum computation). Physicists who, by the nature of their calling, could not help wanting to ask, tried not to. Most of them went on to train their students not to. The most advanced theory in the most fundamental of the sciences was deemed to be stridently contradicting the very existence of truth, explanation, and physical reality.

Not every philosopher abandoned realism. Bertrand Russell and Karl Popper were notable exceptions. Not every physicist did, either. Albert Einstein and David Bohm bucked the trend, and Hugh Everett proposed that physical quantities really do take on more than one value at once (the view we ourselves endorse). On the whole, however, philosophers were uninterested in reality, and although physicists went on using quantum theory to study other areas of physics, research on the nature of quantum processes themselves lost its way.

Things have been gradually improving for a couple of decades, and it has been physics that is dragging philosophy back on track. People want to understand reality, no matter how loudly they may deny that. We are finally sailing past the supposed limits that bad philosophy once taught us to resign ourselves to.

What if the theory is eventually refuted—if some deeper limitation foils the attempt to build a scalable quantum computer? We would be thrilled to see that happen. Such an outcome is by far the most desired one. Not only would it lead to a revision of our fundamental knowledge about physics, we would expect it to provide even more fascinating types of computation. For if something stops quantum mechanics, we shall expect to have an exciting new whatever-stops-quantum-mechanics theory, followed by exciting new whatever-stops-quantum-computers computers. One way or another, there will be no limits on knowledge or progress.

MICHAEL MOYER

Is Space Digital?

FROM *Scientific American*

CRAIG HOGAN BELIEVES that the world is fuzzy. This is not a metaphor. Hogan, a physicist at the University of Chicago and the director of the Fermilab Center for Particle Astrophysics, near Batavia, Illinois, thinks that if we were to peer down at the tiniest subdivisions of space and time, we would find a universe filled with an intrinsic jitter, the busy hum of static. This hum comes not from particles bouncing in and out of being or other kinds of quantum froth that physicists have argued about in the past. Rather, Hogan's noise would come about if space was not, as we have long assumed, smooth and continuous, a glassy backdrop to the dance of fields and particles. Hogan's noise arises if space is made of chunks. Blocks. Bits. Hogan's noise would imply that the universe is digital.

It is a breezy, early autumn afternoon when Hogan takes me to see the machine he is building to pick out this noise. A bright blue shed rises out of the khaki prairie of the Fermilab campus, the only sign of new construction at this forty-five-year-old facility. A fist-wide pipe runs 40 meters from the shed to a long, perpendicular bunker, the former home of a beam that for decades shot subatomic particles north toward Minnesota. The bunker has been reclaimed by what Hogan calls his Holometer, a device designed to amplify the jitter in the fabric of space.

He pulls out a thick piece of sidewalk chalk and begins to write on the side of the cerulean shed, his impromptu lecture detailing how a few lasers bouncing through the tubes can amplify the fine-grain structure of space. He begins by explaining how the two most successful theories of the twentieth century—quantum mechanics and general relativity—cannot possibly be reconciled. At the smallest scales, both break down into gibberish. Yet this same scale seems to be special for another reason: it happens to be intimately connected to the science of information—the 0's and 1's of the universe. Physicists have, over the past couple of decades, uncovered profound insights into how the universe stores information—even going so far as to suggest that information, not matter and energy, constitutes the most basic unit of existence. Information rides on tiny bits; from these bits comes the cosmos.

If we take this line of thinking seriously, Hogan says, we should be able to measure the digital noise of space. Thus he has devised an experiment to explore the buzzing at the universe's most fundamental scales. He will be the first to tell you that it might not work—that he may see nothing at all. His effort is an experiment in the truest sense—a trial, a probe into the unknown. "You cannot take the well-tested physics of spacetime and the well-tested physics of quantum mechanics and calculate what we'll see," Hogan says. "But to me, that's the reason to do the experiment—to go in and see."

And if he does see this jitter? Space and time are not what we thought. "It changes the architecture of physics," Hogan says.

For many years particle physics has not operated on this sort of exploratory model. Scientists spent the late 1960s and early 1970s developing a web of theories and insights that we now know as the Standard Model of particle physics. In the decades since, experiments have tested it with increasing depth and precision. "The pattern has been that the theory community has come up with an idea—for example, the Higgs boson—and you have a model. And the model makes a prediction, and the experiment rules it out or not," Hogan says. Theory comes first, experiments later.

This conservatism exists for a very good reason: particle-physics experiments can be outrageously expensive. The Large Hadron Collider (LHC) at CERN near Geneva required around $5 billion to assemble and currently occupies the attention of thousands of physicists around the world. It is the most sophisticated, complex, and precise machine ever built. Scientists openly wonder if the next generation of particle colliders—at higher energies, larger sizes, and greater expenses—will prove too ambitious. Humanity may simply refuse to pay for it.

A typical experiment at the LHC might include more than 3,000 researchers. At Fermilab, Hogan has assembled a loosely knit team of twenty or so, a figure that includes senior advisers at the Massachusetts Institute of Technology and the University of Michigan who do not participate in day-to-day work at the site. Hogan is primarily a theoretical physicist—largely unfamiliar with the vagaries of vacuum pumps and solid-state lasers—so he has enlisted as coleader Aaron Chou, an experimentalist who happened to arrive at Fermilab at about the same time Hogan was putting his proposal forward. Last summer they were awarded $2 million, which at the LHC would buy you a superconducting magnet and a cup of coffee. The money will fund the entire project. "We don't do any high-tech thing if low-tech will do," Hogan says.

The experiment is so cheap because it is basically an update of the experiment that so famously destroyed the nineteenth century's established wisdom about the backdrop of existence. By the early 1800s, physicists knew that light behaved as a wave. And waves, scientists knew. From a ripple in a pond to sound moving through the air, all waves seemed to share a few essential features. Like sculptures, waves always require a medium—some physical substrate that the waves must travel through. Because light is a wave, the thinking went, it must also require a medium, an invisible substance that permeated the universe. Scientists called this hidden medium the ether.

In 1887 Albert Michelson and Edward Morley designed an experiment that would search for this ether. They set up an interferometer—a device with two arms in the shape of an L that was optimized to measure change. A single source of light would travel the length of both arms, bounce off mirrors at the ends, then recombine where it began. If the length of time it took the light to travel down either arm changed by even a fraction of a microsecond, the recombined light would glow darker. Michelson and Morley set up their interferometer and monitored the light for months as the Earth moved around the sun. Depending on which way the Earth was traveling, the stationary ether should have altered the time it took for the light to bounce down the perpendicular arms. Measure this change, and you have found the ether.

Of course, the experiment found no such thing, thus beginning the destruction of a cosmology hundreds of years old. Yet like a forest obliterated by fire, clearing the ether made it possible for revolutionary new ideas to flourish. Without an ether, light traveled at the same speed no matter how you were moving. Decades later Albert Einstein seized on this insight to derive his theories of relativity.

Hogan's interferometer will search for a backdrop that is much like the ether—an invisible (and possibly imaginary) substrate that permeates the universe. By using two Michelson interferometers stacked on top of each other, he intends to probe the smallest scales in the universe, the distance at which both quantum mechanics and relativity break down—the region where information lives as bits.

The Planck scale is not just small—it is the smallest. If you took a particle and confined it inside a cube less than one Planck length on each side, general relativity says that it would weigh more than a black hole of that same size. But the laws of quantum mechanics say that any black hole smaller than a Planck length must have less than a single quantum of energy, which is impossible. At the Planck length lies paradox.

Yet the Planck length is much more than the space where quantum mechanics and relativity fall apart. In the past few decades an argument over the nature of black holes has revealed a wholly new understanding of the Planck scale. Our best theories may break down there, but in their place something else emerges. The essence of the universe is information, so this line of thinking goes, and the fundamental bits of information that give rise to the universe live on the Planck scale.

"Information means distinctions between things," explained the Stanford University physicist Leonard Susskind during a lecture at New York University last summer. "It is a very basic principle of physics that distinctions never disappear. They might get scrambled or all mixed up, but they never go away." Even after this magazine gets dissolved into pulp at the recycling plant, the information on these pages will be reorganized, not eliminated. In theory, the decay can be reversed—the pulp reconstructed into words and photographs—even if in practice the task appears impossible.

Physicists have long agreed on this principle except in one special case. What if this magazine were to be thrown into a black hole? Nothing can ever emerge from a black hole, after all. Throw these pages into a black hole, and that black hole will appear almost exactly the same as it did before—just a few grams heavier, perhaps. Even after Stephen Hawking showed in 1975 that black holes can radiate away matter and energy (in the form that we now call Hawking radiation), this radiation seemed to be devoid of structure, a flat bleat at the cosmos. He concluded that black holes must destroy information.

Nonsense, argued a number of Hawking's colleagues, among them Susskind and Gerard 't Hooft, a theoretical physicist at Utrecht University in the Netherlands who would go on to win the Nobel Prize. "The whole structure of everything we know would disintegrate if you opened the door even a tiny bit for the notion of information to be lost," Susskind explains.

Hawking was not easily convinced, however, and so over the following two decades physicists developed a new theory that could account for the discrepancy. This is the holographic principle, and it holds that when an object falls into a black hole, the stuff inside may be lost, but the object's information is somehow imprinted onto a surface around the black hole. With the right tools, you could theoretically reconstruct this magazine from a black hole just as you could from the pulp at the recycling plant. The black hole's event horizon—the point of no return—serves double duty as a ledger. Information is not lost.

The principle is more than just an accounting trick. It implies that whereas the world we see around us appears to take place in three dimensions, all the information about it is stored on surfaces that have just two dimensions. What is more, there is a limit to how much information can be stored on a given surface area. If you divide a surface up like a checkerboard, each square two Planck lengths on a side, the information content will always be less than the number of squares.

In a series of papers in 1999 and 2000, Raphael Bousso, now at the University of California, Berkeley, showed how to extend this holographic principle beyond the simple surfaces around black holes. He imagined an object surrounded by flashbulbs popping off in the dark. Light that traveled inward defined a surface—a bubble collapsing at the speed of light. It is on this two-dimensional surface—the so-called light sheet—that all the information about you (or a flu virus or a supernova) is stored.

This light sheet, according to the holographic principle, does a lot of work. It contains information about the position of every particle inside the sheet, every electron and quark and neutrino, and every force that acts on them. Yet it would be wrong to think about the light sheet as a piece of film, passively recording the real stuff that happens out in the world. Instead the light sheet comes first. It projects the information contained on its surface out into the world, creating all that we see. In some interpretations, the light sheet does not just generate all the forces and particles—it gives rise to the fabric of spacetime itself. "I believe that spacetime is what we call emergent," says Herman Verlinde, a physicist at Princeton University and a former student of 't Hooft's. "It will come out of a bunch of 0's and 1's."

One problem: although physicists mostly agree that the holographic principle is true—that information on nearby surfaces contains all the information about the world—they know not how the information is encoded, or how nature processes the 1's and 0's, or how the result of that processing gives rise to the world. They suspect the universe works like a computer—that information conjures up what we perceive to be physical reality—but right now that computer is a big black box.

Ultimately the reason why physicists are so excited about the holographic principle, the reason they spent decades developing it—other than to convince Hawking that he was mistaken, of course—is because it articulates a deep connection between information, matter, and gravity. In the end, the holographic principle could reveal how to reconcile the two tremendously successful yet mutually incompatible pillars of twentieth-century physics: quantum mechanics and general relativity. "The holographic principle is a signpost to quantum gravity," Bousso says, an observation that points the way toward a theory that will supersede our current understanding of the world. "We might need more signposts."

Into all this confusion comes Hogan, with no grand theory of everything, armed with his simple Holometer. But Hogan does not need a grand theory. He does not have to solve all these difficult problems. All he has to do is figure out one fundamental fact: Is the universe a bitlike world or isn't it? If he can do that, he will indeed have produced a signpost—a giant arrow pointing in the direction of a digital universe, and physicists would know which way to go.

According to Hogan, in a bitlike world, space is itself quantum —it emerges from the discrete, quantized bits at the Planck scale. And if it is quantum, it must suffer from the inherent uncertainties of quantum mechanics. It does not sit still, a smooth backdrop to the cosmos. Instead, quantum fluctuations make space bristle and vibrate, shifting the world around with it. "Instead of the universe being this classical, transparent, crystalline-type ether," says Nicholas B. Suntzeff, an astronomer at Texas A&M University, "at a very, very small scale, there are these little foamlike fluctuations. It changes the texture of the universe tremendously."

The trick is getting down to the level of this spacetime foam and measuring it. And here we run into the problem of the Planck length. Hogan's Holometer is an attempt to flank a full-scale assault on the Planck length—a unit so small that measuring it with a conventional experiment (such as a particle accelerator) would involve building a machine the approximate size of the Milky Way.

Back when Michelson and Morley were investigating the (nonexistent) ether, their interferometer measured a tiny change—the change in the speed of light as the Earth moved around the sun— by comparing two light beams that had traveled a reasonably long way. In effect, that distance multiplied the signal. So it is with Hogan's Holometer. His strategy for getting down to the Plank length is to measure the accumulated errors that accrue when dealing with any jittery quantum system.

"If I look at my TV set or my computer monitor, everything looks nice and smooth," Chou says. "But if you look at it close-up, you can see the pixels." As it would be with spacetime. At the level we humans are comfortable with—the scale of people and buildings and microscopes—space appears to be this smooth, continuous thing. We never see a car move down the street by instantaneously leaping from one place to the next as if lit by God's own strobe light.

Yet in Hogan's holographic world, this is exactly what happens. Space is itself discrete—or, in the parlance of our times, "quantized." It emerges out of some deeper system, some fundamentally quantum system that we do not yet understand. "It's a slight cheat because I don't have a theory," Hogan says. "But it's only a first step. I can say to these gravitational theorists, 'You guys figure out how it works.'"

Hogan's holometer is set up much like Michelson and Morley's, if Michelson and Morley had had access to microelectronics and 2-watt lasers. A laser hits a beam splitter that separates the light into two. These beams travel down the two 40-meter-long arms of an L-shaped interferometer, bounce off mirrors at each end, then return to the beam splitter and recombine. Yet instead of measuring the motion of the Earth through the ether, Hogan is measuring any change in the length of the paths as a result of the beam splitter being jostled around on the fabric of space. If, at the Planck scale, spacetime thrashes around like a roiling sea, the beam splitter is the dinghy pitching through the froth. In the time it takes the laser beams to travel out and back through the Holometer, the beam splitter will have jiggled just enough Planck lengths for its motion to be detected.

Of course, you might imagine a lot of reasons why a beam splitter might move a few Planck lengths here and there — the rumbling of a car engine outside the building, for instance, or a stiff Illinois wind shaking the foundations.

Such concerns have bedeviled the scientists behind another interferometry project, the twin Laser Interferometer Gravitational-Wave Observatory (LIGO) detectors outside of Livingston, Louisiana, and Hanford, Washington. These massive experiments were built to observe gravitational waves — the ripples in spacetime that follow cosmic cataclysms such as neutron star collisions. Unfortunately for the LIGO scientists, gravitational waves shake the ground at the same frequency as other not so interesting things — passing trucks and falling trees, for instance. So the detectors have to be completely isolated against noise and vibration. (A proposed wind farm near the Hanford facility caused much consternation among physicists because the mere vibration of the blades would have swamped the detectors with noise.)

The shaking that Hogan is looking for happens much faster — a vibration that jitters back and forth a million times a second. As such, it is not subject to the same noise concerns — only the possible interference from nearby AM radio stations broadcasting at the same frequency. "Nothing moves at that frequency," says Stephan Meyer, a University of Chicago physicist who is working on the Holometer. "If we discover that it's moving anyway, that's one of the things that we'll take as a sure sign" that the jitter is real.

And in the world of particle physics, sure signs can be hard to come by. "This is old-fashioned in a way," Hogan says. "It appeals to this old-fashioned style of physics, which is, 'We're going to go and find out what nature does, without prejudice.'" To illustrate, he likes to tell a parable about the origins of relativity and quantum mechanics. Einstein invented the theory of general relativity by sitting at his desk and working out the mathematics from first principles. There were few experimental quandaries that it solved—indeed, its first real experimental test would not come for years. Quantum mechanics, on the other hand, was imposed on the theorists by the puzzling results of experiments. ("No theorist in his right mind would have invented quantum mechanics unless forced to by data," Hogan says.) Yet it has become the most successful theory in the history of science.

In the same way, theorists have for many years been building beautiful theories such as string theory, although it remains unclear how or if it can ever be tested. Hogan sees the purpose of his Holometer as a way to create the puzzling data that future theorists will have to explain. "Things have been stuck for a long time," he says. "How do you unstick things? Sometimes they get unstuck with an experiment."

SYLVIA A. EARLE

The Sweet Spot in Time

FROM *Virginia Quarterly Review*

AN EYE THE size of my fist peered through the window near the bunk where I lay, half awake, just past midnight on July 20, 2012. A goliath grouper, a fish larger than my desk, swerved past me into the dark sea, its mouth brimming with small fish attracted to the lights of the Aquarius Undersea Laboratory, my home for a week on Conch Reef near Key Largo, Florida. For the tenth time, I was living under the sea, experiencing what I had dreamed of doing as a child, living out a fantasy that had begun with Captain Nemo, Jules Verne, and *Twenty Thousand Leagues Under the Sea.*

One of my five fellow aquanauts, marine biologist Mark Patterson, astutely observed that we were about 20,000 millimeters below the surface—some 60 feet—in a warm, dry suite of rooms, with access to the best swimming pool in the world: the ocean. A rectangular hole in the floor was the entrance to the lab and our exit to the sea beyond, with the pressure inside Aquarius keeping the water from rushing in. For as long as twelve hours a day (or night), scientists who work in Aquarius can explore that richly endowed Florida reef, observe and document the behavior of marine life, conduct experiments, enjoy the perspective of a resident, and make the most of the chance to occupy what is presently the world's only "space station in the sea."

When Jules Verne was born, in 1828, there were about a billion people on Earth, and not one of them had a face mask, radio, television, car, or anything made of plastic, among many other aspects of life we now take for granted. The industrial revolution was well underway, with an explosion of new technologies, new agricultural methods, new means of transportation and communication that led to the greatest era of change in human history. From then to now, population has grown sevenfold, and sustained economic growth has advanced more than tenfold for individual income.

When I was born, in the mid-1930s, Earth's 2 billion people were entering an unprecedented era of prosperity, despite crippling wars and diseases that had taken the lives of millions. Penicillin was not available soon enough to save my brother Don from a fatal ear infection, but revolutionary changes were underway across all aspects of life, from medical care and food production to energy sources and what seemed to be magical new ways to instantly communicate with neighbors nearby as well as with people on the other side of the planet.

As a child, living in a ferment of new discoveries, I thought nothing was beyond the power of human enterprise. With wires, tubes, and ingenuity, my father built our family's first radio—a "crystal set" that brought distant voices and music to our New Jersey farm. We planted new kinds of hybrid corn that yielded far more than seeds collected from the previous year's crop. Prosperity surged through the aftermath of the Great Depression and two world wars, doubling the population in the course of a half-century, reaching 4 billion by 1980. With nearly 7 billion of us today and 2 billion more expected by the middle of this century, our growth as a species appears boundless.

But the benefits have come at a cost. About half of the world's original forests have been consumed, most of them since 1950. Conversion of forests, deserts, marshes, grasslands, and other natural systems for agriculture, cities, and other human purposes has resulted in the loss of thousands—about 75 percent—of the native plants and animals that provide the genetic basis for all agricultural crops and domesticated farm animals. At the same time, a full 90 percent of all large wild fish (and many small kinds as well) have disappeared from the world's oceans, the result of devastating industrial fishing. Entire ecosystems, with their treasury of distinctive plants and animals, have been extinguished, underwater as well as above. Only in times of great natural catastrophe—such as when comets or meteors have collided with our planet—has the way forward been so swiftly and dramatically altered. Never before have such dramatic changes been caused by the actions of a single species.

My goal during the July Aquarius mission was to observe and document changes in a part of the world that I had first witnessed as a young scientist in 1953 and later explored during thousands of hours using the ingenious Self-Contained Underwater Breathing Apparatus—SCUBA—developed in the 1940s and marketed as the Aqua-Lung by the pioneering ocean explorer Jacques Cousteau and a colleague, engineer Émile Gagnan. I experienced my first underwater breath in Florida, and during trips to the Florida Keys, I marveled at the clarity of the ocean and the abundance and diversity of life. Pink conchs plowed trails though seagrass meadows, and schools of colorful fish crowded the branches of elkhorn and staghorn coral. Long, bristly antennae marked the presence of lobsters under ledges and in crevices, and elegantly striped and irrepressibly curious Nassau grouper followed me on most dives, and likely would have continued onto the beach but for the limitations of fins and gills.

Six decades later, I note the difference. The water along the Florida Keys is often gray and murky. The great forests of branching coral are gone. The pink conchs and Nassau grouper are mostly memories—the remaining few protected in U.S. waters, owing to their rarity. With care, there is a chance that these and many other species may recover, but some losses are irrecoverable. I missed meeting, for instance, one of Florida's most charismatic animals, the Caribbean monk seal, a playful St. Bernard–size creature that once lolled on beaches throughout the region, sometimes ranging as far north as Galveston. The last one was sighted in 1952. It is now officially listed as extinct.

During the week on Conch Reef this past July, I saw not even one lobster lurking under the ledges but was pleased to record the actions of numerous large barracuda, which were a constant presence above Aquarius, suspended midwater like sleek gray submarines. Several tarpon swept around the reef, always in motion, much like the dozen or so large permit that danced in glistening circles around the pillars under the lab. A few large black grouper and the awesome goliath made smaller species scatter when they loomed into view. Large barrel sponges and lumps of various hard corals nestled within a carpet of brown and green seaweed. Had I not personally witnessed the Key Largo reefs of fifty years before, I might have regarded this as a fine example of a thriving coral system, oblivious to what is now missing, and unaware that the beautiful, newly arrived lionfish and brilliant yellow cup corals on and near Aquarius are actually exotic species that are rapidly displacing natives.

Scientists using Aquarius over the past twenty years have been able to document the decline of Conch Reef and have gathered information on the complexities of the system that will help us understand what has gone wrong. It will provide insight about what can be done to restore this and other reefs. Globally, about half of the coral reefs that existed when I was a child are gone or are in a state of serious decay. In the Caribbean region, including the Florida Keys, the loss is closer to 80 percent, linked to rising water temperatures and the combined effects of coastal development, overfishing, and pollution.

The United States has been a leader in taking actions to address the decline of the health of natural systems; legislation enacted in the 1970s was aimed at protecting the air, water, coastal zones, endangered species, and marine mammals and—of special significance to the Florida Keys—making possible the designation of marine sanctuaries. In 1990, the year I was appointed chief scientist of the National Oceanic and Atmospheric Administration (NOAA), some 2,900 square nautical miles of waters surrounding the Florida Keys were designated as a marine sanctuary, a multiple-use managed area under NOAA's administration.

NOAA was formed as a special agency of the U.S. government within the Department of Commerce in 1970, the year I first lived underwater. On July 20, 1970, exactly one year after Neil Armstrong and Buzz Aldrin put their footprints on the moon, I emerged from the water after two weeks as a resident of a then-luxuriant reef in Lameshur Bay in the U.S. Virgin Islands—my first experience as an aquanaut. I was participating in part of a two-year program, Tektite I and II, sponsored by various agencies of the U.S. government, including NASA. Skylab and the International Space Station were years away when I successfully applied to be among the nation's first "scientist-aquanauts." At the time, no American women astronauts were allowed to fly; nor, as it turned out, were women thought to be appropriate candidates for becoming aquanauts.

When I proposed to be part of a four-man, one-woman team to study the ecology of reefs where the Tektite lab was situated, there was great consternation, according to Dr. Jim Miller, the head of the project. "We didn't expect women to apply," he explained. "But several did, with qualifications every bit as good as those of the men. We weren't sure how to handle it."

Even in the 1960s, men and women were traveling together on airplanes, sailing on ships, going on camping trips, riding long distances on trains—often using the same bathroom facilities. There were women senators, surgeons, pilots, university presidents, and CEOs of major corporations, but the U.S. government was cautious about having men and women live together for two weeks under the sea, with one bunkroom, one shower, and cameras observing everything that took place, 24/7.

Jim Miller was instrumental in the decision to resolve this issue by forming an all-woman team. He was quoted as saying, "Well, half the fish are female, half the dolphins and whales. I guess we could put up with a few women as aquanauts." Whatever the rationale, I was asked to lead a team of four scientists and one engineer, who had two weeks to conduct research projects and experience what it is like to cook, eat, sleep, take freshwater showers, use microscopes, and work inside quarters that resemble a comfortable trailer parked 50 feet underwater. Then, magically, we could step outside into the ocean for hours of exploration and research, breathing air from a double set of scuba tanks or rebreathers similar to the systems used by space-walking astronauts.

I had logged—before the Tektite Project—more than a thousand hours using scuba and little submarines, but for the first time I began to see fish, sponges, corals, even shrimp as individuals, each one with features as distinctive as those of cats and dogs. I could recognize each of the several gray angelfish that lived near the lab, learned to expect when and where certain parrotfish would tuck in to sleep, and could identify some barracuda by their curious, unique behavior. Each had its place in the complex, thriving communities that made up the reefs, seagrass meadows, and open patches of sand that I explored day and night.

As in New York or London or Singapore, it was obvious that it takes more than buildings to make the system work. The reef has its garbage collectors—sea cucumbers, crabs, goatfish among them. Cleaner fish and tiny shrimp provide critical "medical" services, removing parasites and neatly trimming away dead tissue from wounds. A healthy tension exists among predators and prey, with sharks, grouper, snapper, and other top carnivores keeping those lower down on the food chain in top form—lest they become snacks for those higher up.

None of the women's team had met prior to the mission except Ann Hurley and Alina Szmant, both doctoral candidates at the Scripps Institution of Oceanography. The oceanographer Renate True and I, both biologists, with Margaret Lucas, a graduate student engineer from the University of Delaware, completed the team. Many predicted that we would not get along or that, in more ways than one, we couldn't handle the pressure (two and a half times that experienced at sea level). In fact, we bonded quickly and took seriously the unique opportunity to use the ocean as a laboratory. Our team averaged more time in the water than any others in the program, and despite the foreboding of some, we readily mastered the newly developed rebreather systems that made it possible to dive for as long as six hours with plenty of air to spare.

Our success, we were later told, helped open the way for NASA to accept women as full-fledged astronauts, an unintended but welcome outcome of the Tektite program. Captain George Bond, the warm but gruff navy doctor who pioneered the concept of living underwater in the early 1960s, bluntly admitted that he was among those who opposed my participation, not as a woman, but as a mother of three small children. "But some of the aquanauts were fathers!" I huffed. "I know," he said, reflecting an attitude that transcends logic. "But that's different. A lot of things could have gone wrong, and no one wanted to lose a mother."

Some of the glory associated with astronauts rubbed off on the ten aquanaut teams (nine of them all male) in 1970, with a special media blitz focused on the "aquababes," "aquabelles," or, as one tabloid reported, "the aquanaughties." (We mused about the reaction Apollo astronauts would have had if headlines referred to them as "astrohunks.") As leader of the women's team, I was asked to address Congress, lead a ticker-tape parade in an open limousine with Mayor Richard J. Daley down State Street in Chicago, and be the spokesperson for the project on hundreds of occasions —my baptism by fire as a scientist speaking to public audiences. Nothing I had done before as a scientist had attracted anything more than casual interest. Now millions of people wanted to know what it was like to live under the sea.

I took pleasure in turning questions such as "Did you wear lipstick?" and "Did you use a hair dryer?" into a discourse on the importance of the ocean as our primary source of oxygen, the value of coral reefs, mangroves, and marshes as vital buffers against storms, and the delightful nature of fish, shrimp, lobsters, and crabs alive, swimming in the ocean—not just on plates swimming with lemon slices and butter.

Sometimes I tell young women that I come from another planet, because the world I experienced during the first half of my life is so different from the relatively open attitudes about the role of women now prevalent in this country and around the world. My mother could not vote in the first two national elections held during her lifetime, when men of the same age could. As a graduate student, I was cheerfully told that all the coveted paid teaching assistantships would be given to men "because they would be wasted on women who would just get married and have kids." Social changes have paralleled unprecedented advances in technology that have driven growth and the dissemination of knowledge. More has been learned about the nature of the ocean in the past fifty years, perhaps even the last thirty, than in all of preceding history.

In 1961 President John F. Kennedy noted, "We are just at the threshold of our knowledge of the oceans . . . [This knowledge] is more than a matter of curiosity. Our very survival may hinge upon it." The investments made in the decades that followed have forever altered the understanding of the ocean as the driving force underlying climate, weather, planetary chemistry—and, indeed, our "very survival."

Unknown until the late 1970s was the existence of deep-water hydrothermal vents gushing a hot soup of water, minerals, and microbes and fostering complex communities of creatures, including a previously unknown kingdom of microbes that synthesize food in the absence of sunlight and photosynthesis. No one attained access to the deepest sea until 1960, when two men descended to 35,797 feet (greater than the height of Mount Everest) in the submersible bathyscaphe *Trieste* for a brief glimpse of the deepest point on Earth—the Challenger Deep—in the Mariana Trench near Guam. And no one returned to those depths until March 2012, when the Canadian explorer and film director James Cameron ventured there in his personal one-man submersible.

Technologies that enabled humans to go to the moon and send robots to Mars have given us a vitally important view of Earth from afar—a living blue jewel in a vast universe of unreachable, uninhabitable planets and stars, suspended in a seeming emptiness. On a cell phone or iPad or computer, ten-year-old children can now view Google Earth, zoom from space to their backyard, fly the Grand Canyon, and, starting in 2009, dive into "Google Ocean" to vicariously explore the depths of the sea. New methods of gathering, connecting, evaluating, and communicating data—of measuring change over time and projecting future outcomes based on knowledge no other species has the capacity to acquire—are all causes for hope, but the gains need to be approached with a healthy dose of caution. Even now, with all our advances, less than 5 percent of the ocean has been seen, let alone explored or mapped with the same precision and detail presently available for the moon, Mars, or Jupiter.

The great conservationist Rachel Carson, who summed up what was known about the blue part of the planet in her 1951 book, *The Sea Around Us,* was unaware that continents move around at a stately geological pace or that the greatest mountain chains, deepest valleys, broadest plains, and most of life on Earth are in the ocean. Nor did she appreciate that technological advances developed for wartime applications were being mobilized to find, catch, and market ocean wildlife on an unprecedented scale, reaching distant, deep parts of the ocean no hook or net or trawl had ever touched before.

"Eventually man . . . found his way back to the sea," she wrote. "And yet he has returned to his mother sea only on her terms. He cannot control or change the ocean as, in his brief tenancy of earth, he has subdued and plundered the continents."

In her lifetime—1907 to 1964—she did not, could not, know about the most significant discovery concerning the ocean: it is not too big to fail. Fifty years ago, we could not see limits to what we could put into the ocean or what we could take out. Fifty years into the future, it will be too late to do what is possible right now. We are in a "sweet spot" in time. Never again will there be a better time to take actions that can ensure an enduring place for ourselves within the living systems that sustain us. We are at an unprecedented, pivotal point in history, when the decisions we make in the next ten years will determine the direction of the next ten thousand.

"If I could be anywhere . . . anywhere right now, I would want to be here," croons the singer-songwriter Jackson Browne in a lilting tune he composed in 2010. Why here? Why now? Where would you choose, given the power to live on Mars, slip into the future, or glide back decades, centuries, or even millions or billions of years from our "here on Earth" and the twenty-first century's now?

Some might say anywhere *but* here and now! The world today is at war or on the brink of war. Weapons devised by humans can, in a single stroke, eliminate more people than existed on Earth in 1800. Poverty and hunger haunt hundreds of millions. The world economy is deeply troubled. Diseases are rampant. The natural systems that make life on Earth possible are in sharp decline on land, in the atmosphere above, and in the seas below. Earth's natural fabric of life is in shreds, with consequences that threaten our own existence.

Why not escape to Mars? Or leap back or forward in time?

Before the middle of this century, a few astronauts likely will be setting up housekeeping on the red planet, but they will have to take along a life-support system. Water is scarce, and it is the single nonnegotiable thing all life requires. The atmosphere on Mars is about 95 percent carbon dioxide and therefore lethal to humans (much like Earth's atmosphere in her early years), the temperature averages around –55 degrees centigrade (–67 degrees Fahrenheit). Food, shelter, clothing? Better pack them along and, somehow, find water. Perversely, at the same time some are doing their utmost to make the red planet more like the blue one, a process called "terraforming," we seem to be doing all we can to "Marsiform" Earth. Human actions here on the blue planet have caused the abundance and diversity of life, along with the potable fresh water supply, to decrease, while carbon dioxide is increasing.

Earthlings take for granted that the world is blue, embraced by an ocean that harbors most of the life on the planet, contains 97 percent of the water, drives climate and weather, stabilizes temperature, generates most of the oxygen in the atmosphere, absorbs much of the carbon dioxide, and otherwise tends to hold the planet steady—a friendly place in a universe of inhospitable options.

Owing to more than 2 billion years of microbial photosynthetic activity in the sea and several hundred million years of land-based photosynthesis, Earth's atmosphere now is just right for humans— roughly 21 percent oxygen, 79 percent nitrogen, with trace gases, including just enough carbon dioxide to drive photosynthesis and the continuous production of oxygen and food. Even today, one kind of inconspicuous but enormously abundant sea-dwelling blue-green bacteria, *Prochlorococcus,* churns out 20 percent of the oxygen in the atmosphere, thereby supplying one in every five breaths we take. With other planktonic species, as well as sea-grasses, mangroves, kelps, and thousands of other kinds of algae, ocean organisms do the heavy lifting in terms of taking up carbon dioxide and water via photosynthesis, producing sugar that drives great ocean food chains, and yielding atmospheric oxygen along the way. As much as 70 percent of the air we breathe is produced by underwater life.

Should you choose to go back a billion years on Earth, you would find a planet a lot like Mars except for the abundance of water. Life would be mostly microbial. No trees; no flowers or moss or ferns; no bees or bats or birds—a bleak place compared to the great diversity of life that gradually developed, each and every organism doing its part to shape the barren land and seascapes into an increasingly rich and complex living tapestry.

I have dreamed of being able to see the world as it was a hundred million years ago, to dive into a sea filled with sea stars, urchins, sponges, jellies, corals, seaweeds, sharks, horseshoe crabs, shrimp, and corals. There would be no dolphins or whales, but I might see large aquatic counterparts of terrestrial dinosaurs and giant crustaceans competing with sharks as top ocean predators. Think *Jurassic Park* in the sea!

Humans living on Earth 10,000 years ago were on the brink of an unusually favorable warming period, with massive Northern Hemisphere glaciers diminishing, sea levels rising, the land and sea filled with a rich assemblage of life, including many that served as a source of sustenance for our hunter-gatherer ancestors. During the previous 100,000 years, the ingredients were present for human prosperity, but there were challenging swings in climate, and societies were widely dispersed. While people were endowed with intelligence comparable to that of twenty-first-century humans, they lacked the benefits of thousands of generations of collective knowledge that have allowed subsequent societies to thrive.

People then, as now, shared space with plenty of intelligent animals. Dogs, cats, horses, bonobos, chimpanzees, elephants, dolphins, whales, parrots, albatrosses, octopuses, and certain unusually crafty mantis shrimp come to mind. But none of that intelligent life formed the social bonds that led to the great civilizations humans have built by acquiring and sharing knowledge in ways that enabled succeeding generations to advance.

Two of my friends, Dr. Nancy Knowlton and Dr. Jeremy Jackson, both highly respected marine scientists, are affectionately known as the Doctors Doom and Gloom, and for good reasons. Keen observers, they have witnessed and documented a swift, sharp decline in the world's ocean ecosystems. Some once-common species will likely be extinct soon, no matter what we do. Hundreds of "dead zones," largely resulting from recent land-based pollution, plague coastal regions globally. Enormous "garbage patches" of plastic blight the sea, some sinking to the depths, some cast ashore in great windrows, all destined to be permanent evidence of our carelessness.

There are plenty of reasons for despair. The good news is that half of the coral reefs are still in good shape. Ten percent of the sharks, swordfish, bluefin tuna, grouper, snappers, halibut, and wild salmon are still swimming. Best of all, there is widespread awareness that protection of nature is not a luxury. Rather, it is the key to all past, present, and future prosperity. We may be the planet's worst nightmare, but we are also its best hope.

In January 2012, I sat next to a sixty-one-year-old Laysan albatross, admiring the soft white feathers warming her most recent egg, a small oval of hope nestled in a grassy patch on Midway Island, about halfway across the Pacific Ocean. The artist Wyland, the scientist-photographer Susan Middleton, and I were working with National Park and Fish and Wildlife officials to document the status of the land and surrounding ocean—part of the 140,000-square-mile Papahānaumokuākea National Marine Monument designated by President George W. Bush in 2006. The albatross, named Wisdom, appeared serenely indifferent to our presence when we quietly approached to pay our respects. I marveled at the perils she had survived during six decades, including the first ten or so years before she found a lifetime mate. She learned to fly and navigate over thousands of miles to secure enough small fish and squid to sustain herself and, every other year or so, find her way back to this tiny island and small patch of grass where a voraciously hungry chick waited for special delivery meals.

Like those of us who have roots in the twentieth century and now live in the twenty-first, Wisdom has witnessed a time of unprecedented change. She may wonder about the confusing avalanche of plastic debris, the thousands of miles of drift nets and long lines that bring death to many seabirds, the noise and smells of traffic across the ocean and in the air, all of which she encounters during her months at sea. She may be aware that the world has changed in frightening ways in the course of her lifetime, but she cannot know why, and even if she did know, she would not know what to do to about it. We do know why, and we do know what can be done to secure an enduring place for ourselves within the natural systems that keep us alive.

Making peace with nature is the key.

Early in the twentieth century, President Theodore Roosevelt was among those who led a movement to protect natural areas, watersheds, landscapes, and places of cultural and historic interest as national parks—a concept that Ken Burns called "America's best idea ever." Other nations followed, nearly all adopting the concept, so that such parks now embrace a total area of about 14 percent of the land globally. Presently, less than 2 percent of the ocean is protected, but that soon may change.

In the past, the ocean has not required specific policies to be safe from human actions. Polar regions and the deep ocean have been protected by their inaccessibility, and returning to the same place in the sea has been more art than science until recent technologies made pinpoint navigation possible. Weather forecasting, knowledge of currents, tides, and temperature, and advanced communications now make all parts of the ocean safer than ever before for shipping, fishing, mining, finding and retrieving lost ships, and much more. Technologies as sophisticated as those used to access outer space are being applied to exploit the ocean's deep inner space for oil, gas, minerals, and marine life.

Changing, too, are policies about ocean governance. Historically, property rights and boundaries—and thus protection—have been somewhat easier to establish and manage on land than at sea. Far into the twentieth century, nations claimed jurisdiction over the ocean from the shoreline to just 3 nautical miles—the range of a cannon shot in the 1600s. The Dutch jurist Hugo Grotius in 1609 articulated the widely adopted concept of "freedom of the seas" in international waters as "the common heritage of mankind," with peaceful navigation and open access for one and all to fish and extract minerals and other assets. Even today, nearly half the Earth—the "high seas"—is regarded as a largely unregulated global commons, used by all, protected by none.

Over the years, military conflicts, disputes over fishing rights, and other questions of national interest have led to various international treaties and policies, including those that extend national claims over an Exclusive Economic Zone (EEZ), 200 nautical miles out from a country's coastline. The concept began to take hold in the 1940s and became official as a part of the United Nations Convention on the Law of the Sea in 1982. Although, as of 2012, the United States remains the only industrialized Western nation that has not ratified the Law of the Sea Treaty, this country does claim and respect the EEZ provision. The landmass of the United States covers more than 3.5 million square miles, but the EEZ embraces more than 7 million, the largest square mileage of any nation, essentially doubling the size of the country. Because France has jurisdiction over so many islands around the world, that country is second only to the United States in the size of her submerged claims. Australia, surrounded by ocean, has more submerged area under its jurisdiction than land above.

In the mid-1970s, Australia established the Great Barrier Reef Marine Park Authority, and the United States gave sanctuary status to the historic shipwreck *Monitor*: the first of more than 5,000 ocean areas that have since been designated around the world. Most are small, with only a tiny fraction of 1 percent of all the planet's waters set aside for the protection of marine life. This is far short of the goal of 30 percent we were supposed to reach by 2012, a goal set by the World Parks Congress in Durban, South Africa, in 2003. And clearly it is not enough to maintain the vital functions the ocean delivers to us for "free"—the basic life support that heretofore we have taken for granted.

In 2009, in response to being awarded a TED Prize—$100,000 and the chance to make a wish big enough to "change the world"

—I suggested the following: "I wish you would use all means at your disposal—Films! Expeditions! The web! New submarines!—to create a campaign to ignite public support for a global network of marine protected areas, 'hope spots,' large enough to save and restore the ocean, the blue heart of the planet. How much is large enough? Some say 10 percent, some say 30 percent. You decide: How much of your heart do you want to protect? Whatever it is, a fraction of one percent is not enough."

At the rate we've been going, it will be near the end of the century before we can attain the 30 percent goal targeted by the World Parks Congress. All the same, there is a growing awareness that our fate and the ocean's are closely linked. If the ocean is in trouble, so are we. It is, and we are. But there are reasons to be optimistic.

A global conference in Dubai in December 2011 focused attention on "blue carbon," acknowledging the important role the ocean has in taking carbon dioxide from the air—and the urgent need for greater ocean protection globally. The World Economic Forum in Davos in 2012 for the first time devoted several major sessions to critical ocean issues, and soon thereafter the British publication *The Economist* sponsored a World Oceans Summit in Singapore that brought together leaders of industry, science, technology, and conservation, with an emphasis on the connections between human prosperity and healthy ocean systems.

Ocean issues were prominently on the agenda of the 170 leaders who gathered at the Rio+20 conference in Rio de Janeiro in June 2012, but one of the most pressing topics—developing a process to address the governance of the "high seas"—was tabled for two years. "No one—and everyone—owns the high seas," says Ghislaine Maxwell, founder of the TerraMar Project, a name given to the vast blue global commons. She is encouraging people to sign up for virtual citizenship on the TerraMar website in order to provide a collective voice on behalf of half of the planet.

Dozens of scientists have worked together to produce a sobering report, the Ocean Health Index, which was released in the summer of 2012 and offers a comprehensive system for measuring and monitoring the condition of the world's coastal waters, country by country. Topics such as fisheries, tourism, biodiversity, carbon storage, and economic well-being were considered, and scores were assigned from low to high on a scale of 0 to 100. The lowest ranked is Sierra Leone (36); the highest (86) is Jarvis Island, in the Pacific Remote Islands Marine National Monument. The global score, 60, suggests cause for hope, yet it also demonstrates an urgent need for improvement.

Since my TED wish in 2009, several nations have shown leadership in increasing ocean care. The tone was set in 2006 by two presidents: George W. Bush, who designated major areas in the northwestern Hawaiian Islands and the western Pacific, and Anote Tong, leader of the Pacific island republic of Kiribati, who declared protection that year and in 2008 for 158,000 square miles of ocean surrounding the nation's thirty-three equatorial atolls and islands. Another island nation, the United Kingdom, followed in 2010 with what presently is the world's largest fully protected marine reserve: 225,810 square miles around the Chagos Archipelago in the Indian Ocean. Australia will soon surpass this when it implements a recently announced plan to develop a network of marine reserves covering about one-third of the country's territorial waters. The area will include 386,000 square miles of the Coral Sea bordering the Great Barrier Reef. New Caledonia has announced its intention to create a marine park nearly half the size of India—more than half a million square miles of ocean.

Small island nations—Fiji, Palau, the Marshall Islands, the Gilberts, the Maldives, the Seychelles, the Bahamas, Dominica, the Dominican Republic, and dozens of others—have quite suddenly become "large ocean nations," with a major voice in the politics of ocean management. Some have aligned with Japanese interests in continued exploitation of whales, and many have sold licenses to take fish and minerals for cash and economic assistance. But there is a growing consciousness that protecting the ocean can yield greater and more enduring financial and social benefits than traditional exploitation.

Late in August 2012, I witnessed leaders from sixteen Pacific Island nations gather in the Cook Islands for their forty-third annual meeting to discuss topics of mutual concern, including sea-level rise, the decline of fishing, and the growing dependence on imported fossil fuels and on revenues from tourism. The Cook Islands has a population of 20,000 people who live on fifteen islands that together have a landmass that is slightly larger than that of Washington, D.C. Their ocean mass, however, occupies more than a million square miles. In late August, Henry Puna, the charismatic prime minister of that country, announced the creation of what will soon be the world's largest marine park: 424,000 square miles encompassing most of the southern Cook Islands, an area bigger than France and Germany. He observed, "The marine park will provide the necessary framework to promote sustainable development by balancing economic growth interests such as tourism, fishing, and deep sea mining with conserving core biodiversity in the ocean [. . .] a contribution from the Cook Islands to the well-being of not only our peoples, but also of humanity."

Prior to the meeting of island leaders, I accompanied a small group from Conservation International (CI) for several days of diving around Aitutaki, one of the jewel-like atolls of the Cook Islands. We were thrilled with our repeated encounters with a Napoleon wrasse, a spectacularly ornamented fish as large as the goliath grouper that I had recently observed observing me in the Florida Keys. Valued in Asian markets as a delicacy (especially their large lips), the once-common species is now extremely rare. We were also pleased but sad to see a shark, just one, in a part of the ocean that should have had hundreds. My dive buddy, Greg Stone, the leader of CI's marine program and an ocean policy strategist, has worked closely with President Tong, Prime Minister Puna, and other island leaders to foster a grand vision for an integrated "oceanscape," involving the cooperation of all the island-nations in the region to agree on measures needed to protect their shared ocean assets. The presence or absence of sharks and other large fish is a good indicator of reef productivity. "Healthy reefs need sharks, and sharks need healthy reefs," Greg noted. "Both are worth more alive than dead."

Linking "natural capital"—land and sea—to human prosperity and life itself is an idea whose time has come too late to save Steller's sea cows, Caribbean monk seals, the great auks, and gray whales in the Atlantic Ocean; and it may be too late for many species and systems now on the edge of oblivion. But it is not too late to restore some of the world's damaged reefs, mangroves, and marshes, and to make the blue planet safer, healthier, and more resilient. Lucky us: we are residents of Earth, the sweetest place in the universe at this, the sweetest time.

JOHN PAVLUS

Machines of the Infinite

FROM *Scientific American*

ON A SNOWY day in Princeton, New Jersey, in March 1956, a short, owlish-looking man named Kurt Gödel wrote his last letter to a dying friend. Gödel addressed John von Neumann formally even though the two had known each other for decades as colleagues at the Institute for Advanced Study in Princeton. Both men were mathematical geniuses, instrumental in establishing the United States' scientific and military supremacy in the years after World War II. Now, however, von Neumann had cancer, and there was little that even a genius like Gödel could do except express a few overoptimistic pleasantries and then change the subject:

> Dear Mr. von Neumann:
>
> With the greatest sorrow I have learned of your illness . . . As I hear, in the last months you have undergone a radical treatment and I am happy that this treatment was successful as desired, and that you are now doing better . . .
>
> Since you now, as I hear, are feeling stronger, I would like to allow myself to write you about a mathematical problem, of which your opinion would very much interest me . . .

Gödel's description of this problem is utterly unintelligible to nonmathematicians. (Indeed, he may simply have been trying to take von Neumann's mind off his illness by engaging in an acutely specialized version of small talk.) He wondered how long it would take for a hypothetical machine to spit out answers to a problem. What he concluded sounds like something out of science fiction:

If there really were [such] a machine . . . this would have consequences of the greatest importance. Namely, it would obviously
mean that . . . the mental work of a mathematician concerning Yes-
or-No questions could be completely replaced by a machine.

By "mental work," Gödel didn't mean trivial calculations like
adding 2 and 2. He was talking about the intuitive leaps that mathematicians take to illuminate entirely new areas of knowledge.
Twenty-five years earlier, Gödel's now-famous incompleteness theorems had forever transformed mathematics. Could a machine be
made to churn out similar world-changing insights on demand?

A few weeks after Gödel sent his letter, von Neumann checked
into Walter Reed Army Medical Center in Washington, D.C.,
where he died less than a year later, never having answered his
friend. But the problem would outlive both of them. Now known
as P versus NP, Gödel's question went on to become an organizing
principle of modern computer science. It has spawned an entirely
new area of research called computational complexity theory—
a fusion of mathematics, science, and engineering that seeks to
prove, with total certainty, what computers can and cannot do under realistic conditions.

But P versus NP is about much more than just the plastic-and-
silicon contraptions we call computers. The problem has practical
implications for physics and molecular biology, cryptography, national security, evolution, the limits of mathematics, and perhaps
even the nature of reality. This one question sets the boundaries for what, in theory, we will ever be able to compute. And in
the twenty-first century, the limits of computation look more and
more like the limits of human knowledge itself.

The Bet

Michael Sipser was only a graduate student, but he knew someone would solve the P versus NP problem soon. He even thought
he might be the one to do it. It was the fall of 1975, and he was
discussing the problem with Leonard Adleman, a fellow graduate
student in the computer science department at the University of
California, Berkeley. "I had a fascination with P versus NP, had
this feeling that I was somehow able to understand it in a way that
went beyond the way everyone else seemed to be approaching it,"

says Sipser, who is now head of the mathematics department at the Massachusetts Institute of Technology. He was so sure of himself that he made a wager that day with Adleman: P versus NP would be solved by the end of the twentieth century, if not sooner. The terms: one ounce of pure gold.

Sipser's bet made a kind of poetic sense, because P versus NP is itself a problem about how quickly other problems can be solved. Sometimes simply following a checklist of steps will get you to the end result in relatively short order. Think of grocery shopping: you tick off the items one by one until you reach the end of the list. Complexity theorists label these problems P, for "polynomial time," which is a mathematically precise way of saying that no matter how long the grocery list becomes, the amount of time that it will take to tick off all the items will never grow at an unmanageable rate.

In contrast, many more problems may or may not be practical to solve by simply ticking off items on a list, but checking the solution is easy. A jigsaw puzzle is a good example: even though it may take effort to put together, you can recognize the right solution just by looking at it. Complexity theorists call these quickly checkable, "jigsaw puzzle–like" problems NP.

Four years before Sipser made his bet, a mathematician named Stephen Cook had proved that these two kinds of problems are related: every quickly solvable P problem is also a quickly checkable NP problem. The P versus NP question that emerged from Cook's insight—and that has hung over the field ever since—asks if the reverse is also true: Are all quickly checkable problems quickly solvable as well? Intuitively speaking, the answer seems to be no. Recognizing a solved jigsaw puzzle ("Hey, you got it!") is hardly the same thing as doing all the work to find the solution. In other words, P does not seem to equal NP.

What fascinated Sipser was that nobody had been able to mathematically *prove* this seemingly obvious observation. And without a proof, a chance remained, however unlikely or strange, that all NP problems might actually be P problems in disguise. P and NP might be equal—and because computers can make short work of any problem in P, P equals NP would imply that computers' problem-solving powers are vastly greater than we ever imagined. They would be exactly what Gödel described in his letter to von Neumann: mechanical oracles that could efficiently answer just about any question put to them, so long as they could be programmed to verify the solution.

Sipser knew that this outcome was vanishingly improbable. Yet proving the opposite, much likelier, case—that P is not equal to NP—would be just as groundbreaking.

Like Gödel's incompleteness theorems, which revealed that mathematics must contain true but unprovable propositions, a proof showing that P does not equal NP would expose an objective truth concerning the limitations of knowledge. Solving a jigsaw puzzle and recognizing that one is solved are two fundamentally different things, and there are no shortcuts to knowledge, no matter how powerful our computers get.

Proving a negative is always difficult, but Gödel had done it. So to Sipser, making his bet with Adleman, twenty-five years seemed like more than enough time to get the job done. If he couldn't prove that P did not equal NP himself, someone else would. And he would still be one ounce of gold richer.

Complicated Fast

Adleman shared Sipser's fascination, if not his confidence, because of one cryptic mathematical clue. Cook's paper establishing that P problems are all NP had also proved the existence of a special kind of quickly checkable type of problem called NP-complete. These problems act like a set of magic keys: if you find a fast algorithm for solving one of them, that algorithm will also unlock the solution to every other NP problem and prove that P equals NP.

There was just one catch: NP-complete problems are among the hardest anyone in computer science had ever seen. And once discovered, they began turning up everywhere. Soon after Cook's paper appeared, one of Adleman's mentors at Berkeley, Richard M. Karp, published a landmark study showing that twenty-one classic computational problems were all NP-complete. Dozens, then hundreds, soon followed. "It was like pulling a finger out of a dike," Adleman says. Scheduling air travel, packing moving boxes into a truck, solving a sudoku puzzle, designing a computer chip, seating guests at a wedding reception, playing Tetris, and thousands of other practical, real-world problems have been proved to be NP-complete.

How could this tantalizing key to solving P versus NP seem so commonplace and so uncrackable at the same time? "That's why I was interested in studying the P versus NP problem," says Adleman, who is now a professor at the University of Southern California. "The power and breadth of these computational questions just seemed deeply awesome. But we certainly didn't understand them. And it didn't seem like we would be understanding them any time soon." (Adleman's pessimism about P versus NP led to a world-changing invention: a few years after making his bet, Adleman and his colleagues Ronald Rivest and Adi Shamir exploited the seeming incommensurability of P and NP to create their eponymous RSA encryption algorithm, which remains in wide use for online banking, communications, and national security applications.)

NP-complete problems are hard because they get complicated fast. Imagine you are a backpacker planning a trip through a number of cities in Europe, and you want a route that takes you through each city while minimizing the total distance you will need to travel. How do you find the best route? The simplest method is just to try out each possibility. With five cities to visit, you need to check only twelve possible routes. With ten cities, the number of possible routes mushrooms to more than 180,000. At sixty cities, the number of paths exceeds the number of atoms in the known universe. This computational nightmare is known as the traveling-salesman problem, and in over eighty years of intense study, no one has ever found a general way to solve it that works better than trying every possibility one at a time.

That is the perverse essence of NP-completeness—and of P versus NP: not only are all NP-complete problems equally impossible to solve except in the simplest cases—even if your computer has more memory than God and the entire lifetime of the universe to work with—they seem to pop up everywhere. In fact, these NP-complete problems don't just frustrate computer scientists. They seem to put limits on the capabilities of nature itself.

Nature's Code

The pioneering Dutch programmer Edsger Dijkstra understood that computational questions have implications beyond mathematics. He once remarked that "computer science is no more about computers than astronomy is about telescopes." In other words, computation is a behavior exhibited by many systems besides those made by Google and Intel. Indeed, any system that transforms inputs into outputs by a set of discrete rules — including those studied by biologists and physicists — can be said to be computing.

In 1994 the mathematician Peter Shor proved that cleverly arranged subatomic particles could break modern encryption schemes. In 2002 Adleman used strands of DNA to find an optimal solution to an instance of the traveling-salesman problem. And in 2005 Scott Aaronson, an expert in quantum computing who is now at MIT's Computer Science and Artificial Intelligence Laboratory, used soap bubbles, of all things, to efficiently compute optimal solutions to a problem known as the Steiner tree. These are all exactly the kinds of NP problems that computers should choke their circuit boards on. Do these natural systems know something about P versus NP that computers don't?

"Of course not," Aaronson says. His soap bubble experiment was actually a reductio ad absurdum of the claim that simple physical systems can somehow transcend the differences between P and NP problems. Although the soap bubbles did "compute" perfect solutions to the minimum Steiner tree in a few instances, they quickly failed as the size of the problem increased, just as a computer would. Adleman's DNA-strand experiment hit the same wall. Shor's quantum algorithm does work in all instances, but the factoring problem that it cracks is almost certainly not NP-complete. Therefore the algorithm doesn't provide the key that would unlock every other NP problem. Biology, classical physics, and quantum systems all seem to support the idea that NP-complete problems have no shortcuts. And that would be true only if P did not equal NP.

"Of course, we still can't prove it with airtight certainty," Aaronson says. "But if we were physicists instead of complexity theorists, 'P does not equal NP' would have been declared a law of nature long ago — just like the fact that nothing can go faster than the speed of light." Indeed, some physical theories about the fundamental nature of the universe — such as the holographic principle, suggested by Stephen Hawking's work on black holes — imply that the fabric of reality itself is not continuous but made of discrete bits, just like a computer. Therefore, the apparent intractability of

NP problems—and the limitations on knowledge that this implies —may be baked into the universe at the most fundamental level.

Brain Machine

So if the very universe itself is beholden to the computational limits imposed by P versus NP, how can it be that NP-complete problems seem to get solved all the time—even in instances where finding these solutions should take trillions of years or more?

For example, as a human fetus gestates in the womb, its brain wires itself up out of billions of individual neurons. Finding the best arrangement of these cells is an NP-complete problem—one that evolution appears to have solved. "When a neuron reaches out from one point to get to a whole bunch of other synapse points, it's basically a graph-optimization problem, which is NP-hard," says evolutionary neurobiologist Mark Changizi. Yet the brain doesn't actually solve the problem—it makes a close approximation. (In practice, the neurons consistently get within 3 percent of the optimal arrangement.) The *Caenorhabditis elegans* worm, which has only 302 neurons, still doesn't have a perfectly optimal neural-wiring diagram, despite billions on billions of generations of natural selection acting on the problem. "Evolution is constrained by P versus NP," Changizi says, "but it works anyway, because life doesn't always require perfection to function well."

And neither, it turns out, do computers. That modern computers can do anything useful at all—much less achieve the wondrous feats we all take for granted on our video-game consoles and smartphones—is proof that the problems in P encompass a great many of our computing needs. For the rest, often an imperfect approximating algorithm is good enough. In fact, these "good enough" algorithms can solve immensely complex search and pattern-matching problems, many of which are technically NP-complete. These solutions are not always mathematically optimal in every case, but that doesn't mean they aren't useful.

Take Google, for instance. Many complexity researchers consider NP problems to be, in essence, search problems. But according to Google's director of research, Peter Norvig, the company takes pains to avoid dealing with NP problems altogether. "Our users care about speed more than perfection," he says. Instead Google researchers optimize their algorithms for an even faster computational complexity category than P (referred to as linear time) so that search results appear nearly instantaneously. And if a problem comes up that cannot be solved in this way? "We either reframe it to be easier, or we don't bother," Norvig says.

That is the legacy and the irony of P versus NP. Writing to von Neumann in 1956, Gödel thought the problem held the promise of a future filled with infallible reasoning machines capable of replacing "the mental work of a mathematician" and churning out bold new truths at the push of a button. Instead, decades of studying P versus NP have helped build a world in which we extend our machines' problem-solving powers by embracing their limitations. Lifelike approximation, not mechanical perfection, is how Google's autonomous cars can drive themselves on crowded Las Vegas freeways and IBM's Watson can guess its way to victory on *Jeopardy!*

Gold Rush

The year 2000 came and went, and Sipser mailed Adleman his ounce of gold. "I think he wanted it to be embedded in a cube of Lucite, so he could put it on his desk or something," Sipser says. "I didn't do that." That same year the Clay Mathematics Institute in Cambridge, Massachusetts, offered a new bounty for solving P versus NP: $1 million. The prize helped to raise the problem's profile, but it also attracted the attention of amateurs and cranks; nowadays, like many prominent complexity theorists, Sipser says, he regularly receives unsolicited e-mails asking him to review some new attempt to prove that P does not equal NP—or worse, the opposite.

Although P versus NP remains unsolved, many complexity researchers still think it will yield someday. "I never really gave up on it," Sipser says. He claims to still pull out pencil and paper from time to time and work on it—almost for recreation, like a dog chewing on a favorite bone. P versus NP is, after all, an NP problem itself: the only way to find the answer is to keep searching. And while that answer may never come, if it does, we will know it when we see it.

MICHELLE NIJHUIS

Which Species Will Live?

FROM *Scientific American*

THE ASHY STORM petrel, a tiny, dark gray seabird, nests on eleven rocky, isolated islands in the Pacific Ocean off the coasts of California and Mexico. Weighing little more than a hefty greeting card and forced to contend with invasive rats, mice, and cats; aggressive seagulls; oil spills; and sea-level rise, it faces an outsize fight for survival. At last count, only 10,000 remained. Several other species of storm petrels are similarly endangered.

Yet at least one conservation group has decided to ignore the petrel. In the winter of 2008 the Wildlife Conservation Society was focusing its far-flung efforts on a small number of animals. The society's researchers had spent months analyzing thousands of declining bird and mammal species around the world and had chosen several hundred that could serve as cornerstones for the organization's work. They then turned to people with decades of experience studying wildlife to further narrow the possibilities.

Dozens of these experts gathered in small conference rooms in New York City, southwestern Montana, and Buenos Aires to make their choices. They judged each species for its importance to its ecosystem, its economic and cultural value, and its potential to serve as a conservation emblem. They voted on each animal publicly, holding up red, yellow, or green cards. When significant disagreement occurred, the experts backed up their reasoning with citations, and the panels voted again. By the middle of the first day most panels had eliminated more than half the species from their lists.

At some point in the afternoon, however, in every meeting, the reality of the process would hit. As entire groups of species, including storm petrels, were deemed valuable but not valuable enough, a scientist would quietly shut down, shoulders slumped and eyes glazed. "I'm just overwhelmed," he or she might say. Panel members would encourage their colleague, reminding him or her that these choices were necessary and that the science behind them was solid. John Fraser, a conservation psychologist who moderated the panels, would suggest a coffee break. "I'd say, 'I'm sorry, but we have to stop. This is a very important part of the process,'" he remembers. "It was important to recognize the enormity of what we were doing—that we were confronting loss on a huge scale."

The experts knew that all conservation groups and government agencies were coping with similar choices in tacit ways, but the Wildlife Conservation Society process made those decisions more explicit and more painful. As budgets shrink, environmental stresses grow, and politicians and regulators increasingly favor helping the economy over helping the planet, many scientists have come to acknowledge the need for triage. It is time, they say, to hold up their cards.

Triage: A Four-Letter Word

The concept of conservation triage is based loosely on medical triage, a decision-making system used by battlefield medics since the Napoleonic Wars. Medical triage has several variations, but all of them involve sorting patients for treatment in difficult situations where time, expertise, or supplies, or all three, are scarce. The decisions are agonizing but are considered essential for the greater good.

In 1973, however, when the U.S. Congress passed the Endangered Species Act, the mood was not one of scarcity but of generosity. The act, still considered the most powerful environmental law in the world, stipulated eligibility for protection for all non-pest species, from bald eagles to beetles. Later court decisions confirmed its broad reach. In their book *Noah's Choice*, the journalist Charles C. Mann and the economist Mark L. Plummer describe the act's reasoning as the Noah Principle: all species are fundamentally equal, and everything can and should be saved, regardless of its importance to humans.

Trouble arose in the late 1980s, when proposed endangered-species listings of the northern spotted owl and some salmon varieties threatened the economic interests of powerful timber and fishing industries, setting off a series of political and legal attempts to weaken the law. Environmentalists fought off the attacks, but the bitter struggle made many supporters suspicious of any proposed changes to the law, even those intended to increase its effectiveness. In particular, proponents feared that any overt attempt to prioritize endangered species—to apply the general principle of triage—would only strengthen opponents' efforts to try to cut species from the list. If such decisions had to happen, better that they be made quietly, out of political reach.

"The environmental community was always unwilling to talk about triage," says Holly Doremus, a law professor at the University of California, Berkeley. "Even though they knew it was going on, they were unwilling to talk about it."

Today triage is one of the most provocative ideas in conservation. To many, it invokes not only political threats to laws such as the Endangered Species Act but an abandonment of the moral responsibility for nature implied in the Noah Principle. "Triage is a four-letter word," conservation biologist Stuart Pimm recently told *Slate*'s Green Lantern blog. "And I know how to count."

Pine Trees or Camels

Conservationists who are pushing for explicit triage say they are bringing more systematic thinking and transparency to practices that have been carried out implicitly for a long time. "The way we're doing it right now in the United States is the worst of all possible choices," says Tim Male, a vice president at Defenders of Wildlife. "It essentially reflects completely ad hoc prioritization." Politically controversial species attract more funding, he says, as do species in heavily studied places: "We live in a world of unconscious triage."

In recent years researchers have proposed several ways to make triage decisions, with the aim of providing maximum benefit for nature as a whole. Some scientists argue for weighting species according to their role in the ecosystem, an approach we might call "function first." Threatened species with a unique job, they say, or "umbrella" species whose own survival ensures the survival of many others should be protected before those with a so-called re-

dundant role. One example is the campaign to protect the Rocky
Mountains' high-elevation whitebark pines, trees stressed by warming temperatures and associated beetle outbreaks. Because high-
fat whitebark pine nuts are an important food source for grizzly
bears in the fall and spring, many conservation groups view the
pine as a priority species.

The advantage of this function-first approach is that it focuses
on specific ecological roles rather than raw numbers of species,
giving conservationists a better chance at protecting functioning
ecosystems. The approach, however, is useful only in well-under-
stood systems, and the number of those is small. An exclusively
function-first analysis would almost certainly leave many ecologically important species behind.

As an alternative, the EDGE (Evolutionarily Distinct and Globally Endangered) of Existence program run by the Zoological Society of London argues for prioritizing species at the genomic level,
an approach we might call "evolution first." Rather than focusing
on well-known species with many near relatives, the EDGE program favors the most genetically unusual threatened species. Examples include the two-humped Bactrian camel; the long-beaked
echidna, a short, spiny mammal that lays eggs; and the Chinese
giant salamander, which can grow to six feet in length.

The evolution-first approach emphasizes the preservation of genetic diversity, which can help all the world's species survive and
adapt in fast-changing environmental conditions by providing a
robust gene pool. But as the University of Washington ecologist
Martha Groom points out, exclusive use of the approach could
miss broader threats that affect entire taxa, leaving groups of species vulnerable to wholesale extinction. "What if a whole branch
of the evolutionary tree is endangered?" she asks. "What do we do
then?"

Of course, species are valuable for many different reasons.
Some play a vital role in the ecosystem, some have unique genes,
some provide extensive services to humans. No single criterion
can capture all these qualities. The Wildlife Conservation Society
combined different triage approaches in its analyses: it gave priority to threatened species that have larger body size and wider geographic range, reasoning that protection of these creatures would
likely benefit many other plants and animals. It also gave higher

rankings to species with greater genetic distinctiveness. The expert panels then considered more subjective qualities, such as cultural importance and charisma, which, like it or not, are important to fundraising.

Groom, who helped to lead the society's analysis, says it opted for the combined approach because much of the information she and her colleagues needed was unknown or unquantifiable. "There's an awful lot of uncertainty and ignorance about all species," she says. But with a combination of available data and expert opinions, the analysis identified a small group of "global priority" species that the organization can focus on.

Ecosystems over Species

Given the importance of protecting not simply individual animals but also the relations among them, some researchers say that triage approaches should select among ecosystems instead of species. In the late 1980s the British environmentalist Norman Meyers proposed that his global colleagues try to protect the maximum number of species by focusing on land areas that were full of plants found nowhere else on the planet and that were also under pressing environmental threats.

Meyers called such places hotspots. He and his partners at Conservation International eventually identified twenty-five hotspots worldwide, from coastal California to Madagascar, that they thought should top priority lists. In a sense, the approach combines the function-first and evolution-first processes: it protects ecological relations by focusing on entire ecosystems, and it protects genetic diversity by prioritizing endemic species. The idea caught on, and it influences decisions by many philanthropists, environmental organizations, and governments today.

Nevertheless, in recent years researchers have criticized hotspots for oversimplifying a global problem and for giving short shrift to human needs. "It was brilliant for its time," says Hugh Possingham of the University of Queensland in Australia. "But it used just two criteria."

In an effort to refine the concept, Possingham and his colleagues developed Marxan, a software program that is now in wide use. It aims to maximize the effectiveness of conservation reserves by considering not only the presence of endemic species and the level of conservation threats but also factors such as the cost of protection and "complementarity"—the contribution of each new reserve to existing biodiversity protections. Mangrove forests, for instance, are not particularly rich in species and might never be selected by a traditional hotspot analysis; Possingham's program, however, might recommend protection of mangrove forests in an area where representative swaths of other, more diverse forest types had already been preserved, resulting in a higher total number of species protected.

Protected areas and parks, however, can be difficult to establish and police, and because climate change is already shifting species ranges, static boundaries may not offer the best long-term protection for some species. In response, Possingham has created a resource-allocation process that goes well beyond the selection of hotspots, allowing decision makers to weigh costs, benefits, and the likelihood of success as they decide among different conservation tactics. "You do actions—you don't do species," Possingham says. "All prioritizations should be about actions, not least because in many cases actions help multiple species."

The New Zealand Department of Conservation has used the resource-allocation process to analyze protection strategies for about 710 declining native species. It concluded that by focusing on the actions that were cheapest and most likely to succeed, it could save roughly half again as many plants and animals from extinction with the same amount of money. Although some scientists worry that the process places too much emphasis on preserving sheer numbers of threatened species and too little on preserving ecosystem function, resource-allocation analysis is now underway in Australia, and Possingham has spoken with U.S. Fish and Wildlife Service officials about the process.

"People think triage is about abandoning species or admitting defeat," says Madeleine Bottrill of Conservation International, who is a colleague of Possingham's. To the contrary, she argues: by quantifying the costs and payoffs of particular actions, the trade-offs become explicit. Agencies and organizations can identify what is being saved, what is being lost, and what could be saved with a bigger budget, giving them a much stronger case for more funding.

Success Breeds Success

It is possible that the very act of setting priorities more overtly could inspire societies to spend more money on conservation efforts. Defenders of Wildlife's Tim Male says prioritization schemes, far from exposing nature to political risks, offer practical and political advantages. "If we focus more effort on the things we know how to help, we're going to produce more successes," he says. "More successes are a really compelling argument—not just to politicians but to ordinary people—for why [conservation programs] should continue."

Trailing behind such successes, however, are undeniable losses, and true triage must acknowledge them. "We're very good as humans, aren't we, at justifying any amount of work on anything based on undeclared values," says Richard Maloney of the New Zealand Department of Conservation. "We're not very good at saying, 'Because I'm working on this species, I'm not going to fund or work on these seven or eight species, and they're going to go extinct.'" And yet Maloney himself is reluctant to name the species likely to lose out in his agency's resource-allocation analysis. Rockhopper penguins—whose vital supply of krill has declined because of shrinking sea ice driven by climate change—fall to the bottom of the department's list because of the costly, long-shot measures needed to protect them. Yet the species's low priority, Maloney argues, should be seen not as a death sentence but as a call to action by other groups.

Sooner or later, though, a vulnerable species or habitat—the rockhopper penguin, the whitebark pine ecosystem—will require measures too expensive for any government or group to shoulder. What then? Do societies continue to pour money into a doomed cause or allow a species to die out, one by one, in plain sight? Even though the conversation about triage has come a long way, many conservationists remain uncomfortable taking responsibility for the final, fateful decisions that triage requires.

The central difficulty is that, just as with battlefield triage, the line between opportunity and lost cause is almost never clear. In the 1980s, when the population of California condors stood at just 22, even some environmentalists argued that the species should be permitted to "die with dignity." Yet others made an evolution-first argument, calling for heroic measures to save the rare Pleistocene relic. With heavy investments of money, time, and expertise, condors were bred in captivity and eventually returned to the wild, where 217 fly today, still endangered but very much alive.

"We can prevent extinction; we've demonstrated that," says John Nagle, a law professor at the University of Notre Dame who has written extensively about environmental issues. But "knowing that an extinction was something we could have stopped and chose not to—I think that's where people kind of gulp and don't want to go down that road," he adds.

Similarly, by creating what the prominent restoration ecologist Richard Hobbs calls a "too-hard basket" for species that would cost too much to save, a triage system could allow societies to prematurely jettison tough cases, choosing short-term economic rewards over long-term conservation goals. The Endangered Species Act itself has one provision for such a too-hard basket—it allows for a panel of experts that can, in unusual circumstances, permit a federal agency to violate the act's protections. But the so-called God Squad is deliberately difficult to convene and has so far made only one meaningful exemption to the act: letting the Forest Service approve some timber sales in habitats of the struggling northern spotted owl.

As climate change, population expansion, and other global pressures on biodiversity continue, however, more and more species are likely to require heroic measures for survival. Prioritizing species by ecological function, evolutionary history, or other criteria will help shape conservation strategies, but for the greater good of many other species, societies will almost certainly have to consciously forgo some of the most expensive and least promising rescue efforts.

In the United States, legal scholars have suggested ways of reforming the Endangered Species Act to reckon with this reality—to help the law bend instead of break under political pressure. Yet Nagle says that the essence of the law, the Noah Principle, remains acutely relevant. Given the temptations that accompany triage, he says, the exhortation to save all species remains a worthy, and perhaps even necessary, goal. Just as a battlefield medic works unstintingly to save lives, even while knowing that he or she cannot save them all, societies should still aspire to the Noah Principle—and stuff the ark to the brim.

RICK BASS

The Larch

FROM *Orion*

FOR AS LONG as I can remember having known them, I have
been wanting to write about larch trees. I've been putting it off for
fifteen years because, for one thing, it's like writing about lichens
or a clock that moves its hour hand only a fraction of an inch each
year. (During my fifteen years of procrastination, one of the old
giants has perhaps added only half an inch to its girth. And yet,
magnified throughout the forest—millions of such inches—surely
the power of glaciers has been equaled.) But also, I've simply been
afraid of attempting an essay about the larch, such is the reverence
I have for the tree.

They don't speak, not even in the wind, really—unlike the
soughing and clacking limbs and trunks of the limber lodgepole
and the playing-card deck-shuffling clatter of aspen leaves in summer and fall—and even their dying comes slow. Sometimes a big
larch will remain upright for a hundred years or longer after it's
died—perishing in a huge fire or, occasionally, just dying, and finally rotting—and even after they fall over, snapping the other
trees around them on their way and shaking the earth with their
thunder, they remain there, solid and real, for centuries, and in
many ways as alive or more so in their decomposition—possessing,
or housing, more writhing life in that rotting than they did even in
the upright living days of green and gold.

They are, of course, every bit as glorious in life as in death.
While among the green and the living, they possess numerous
attributes, one of the most underrated of which is that of water
pump, intercepting snowmelt and surface sheet flow that might

otherwise drain off to the nearest road and be carried away from the forest, unutilized. But the larch capture and claim and hold within the forest that water, and they convert it to astounding height and magnificent breadth.

What else is the function of a forest, first and foremost, if not a place to do this: to capture and filter water and merge with sunlight, to create intricate being, intricate matter?

The big larches don't just claim and hold that runaway water; they circulate it, too, each tree a miniature weather system unto itself, returning hundreds of gallons of water to the ecosystem each day in the form of transpiration, a fine, even invisible mist emanating from the needles, just as lung-damp breath is emitted from a man or a woman; and on cold and damp mornings, you can see the same clouds of steam rising in plumes from the larch trees as you would see sifting from the mouths and nostrils of a forest of men and women.

This is not to say that the larch are gluttons, greedsome water scavengers totally out of control. One of the reasons they can get so big is that they can live so damn long if you let them—if you don't saw them down. Around the age of two hundred or three hundred, they really begin to hit their stride, and having clearly gained a secure place in the canopy, they can concentrate their efforts almost exclusively thereafter on getting roly-poly big around the middle; in the Yaak Valley, where I live, there are larches that have lived to be six and seven hundred years old.

They can prosper with either seasonal or steady access to water, though they can prosper also on the drier sites, such as those favored by the ponderosa pine. When need be, they can be prim and frugal with water, as in a drought, calibrating their internal balances with exquisite deftness to slow their growth as if almost into dormancy, where they hunker and lurk, giant and calm, awaiting only the freedom, the release, of the next wet cycle.

And they tolerate—flourish in, actually—fire, about which I will say more later.

Like any tree, they have certain diseases that can compromise their species—dwarf mistletoe, which sometimes weakens them through parasitic attrition, and larch casebearer beetle, which is kept in check by the fires and by the incredible battalions of flickers and woodpeckers (pileated, black-backed, Lewis's, downy, hairy, northern three-toed, and more) that sweep and swoop through these forests, drilling and ratting and tatting and pounding, searching and probing and pecking and cleaning and aerating almost ceaselessly during the growing months. But for the most part, the larch remain relatively secure in a world where so many other trees—fir, spruce, dogwood, oak, pine—are undergoing an epidemic of rot and beetles and blight and gypsy moths and acid rain.

Who can say for sure why the great larches are—for now—weathering the howling world so well, in these decades of such intense environmental degradation. On a purely intuitive level, I suspect that the answer has something to do with the larch's ancient jurisprudence—with the way it has evolved so carefully, so precisely, so uniquely and specifically to be safe in the world.

The larch is two things, not one: a deciduous conifer, bearing its seeds in cones but losing its needles each autumn, and it has selected the best attributes of each—the ancient conifers and the more recent deciduous trees—to fit into the one place on Earth that would most have it, the strange dark cant of the Yaak, tipped into a magic seam between the northern Rockies and the Pacific Northwest.

Or perhaps their sturdiness, their calm and elegant forbearance in a world filled with drought and fire and disease, comes not from their wise evolutionary strategy of keeping one foot in each world, but from the fact that they lie so extraordinarily low, sleeping or near dormant for the eight or nine months of the year when they either have no needles at all (the first little spindly paintbrush nubs not sprouting out some years until May) or their needles have already shut down production and have begun to turn bright autumn gold, which can happen as early as August. Perhaps, by sleeping so much, they age only one year to other trees' two or three or even four years.

In this regard, they are like a super-aspen, or a super-oak, calibrating their explosive leap of life to reside perfectly within that tipped, thin window of sunlight and moisture in the Yaak, the three-month growing season, and then shedding their needles, just as the oak and aspen drop their leaves, once that period of growth has ended, for there's no need to invest in keeping them, dormant or barely alive, through the winter. Better to shut it all down and sleep completely.

But the larch are like a super-pine, too, or a super-fir, possessing the eager colonizing tricks of the conifers that have flourished for the last eon in the huge, landscape-altering sweeps of drama that follow the large fires in the northern Rockies—casting their seed-sprung cones from high above, down into the fertile ash, and in that way stretching like a living wave, or like an animal walking into new territory.

(In the northern Rockies, some things run from a fire and other things follow it—elk following the green grass that follows the autumn-before's flames, so that in one sense, perhaps as seen through squinted eyes, the elk can be said to be the grass can be said to be the fire, with very little difference in the movements of any of the three of them—all three generated and directed by the same force—and to that series of waves can be added the larch, colonizing those new burns and then reaching for the sky, rising slowly into 150-foot peaks that can take centuries to crest.)

So the larch, like the Yaak itself, is two things, not one: fire and rot, shadow and light. And in keeping with another of the stories of the Yaak—the fact that what is rare or even vanished from much of the rest of the world is often still present, sometimes in abundance, here—the larch is the rarest form of old growth in the West, though in the Yaak it is the most common form.

Biologist Chris Filardi has looked at the maps of distribution for larch, as well as the habitat type found here, and has declared that the Yaak is "the epicenter of larch." This species is the one thing, I think, that is most truly ours. So many of the Yaak's other wonders are down to nearly the thin edge of nothing—five or six wolves, fewer than twenty grizzlies, a handful of lynx, a dozen mated pairs of bull trout, one occasional woodland caribou, a handful of wolverines, fourteen little roadless areas, one pure population of inland redband trout . . .

The larch, however, is at the edge of nothing. This is the center of the center. Increasingly, I am convinced that the larch trees possess, more than any other single thing, the spirit of the Yaak.

Their interior wood, all the way through, is the red-orange color of campfire coals, a darker orange than a pumpkin, darker orange than the fur of an elk, and while I haven't found a scientist yet who can or will dare guess why the inside of the tree, never seen except when the tops snap off, or when the saw bisects the flesh, should be that firesome color, you would not be able to disprove, I think, the notion that there might be some distant parallel pattern or connection, out at or beyond the edge of our present knowledge, wherein fire likes, and is drawn to, the color of the larch's interior; for the larch is nothing if not birthed of fire.

And again, not just any fire, but the strangeness here, in the Yaak, of fire sweeping through and across a lush and rainy land that, when it is not burning, is rotting, and which is always, even in the rotting and the burning, growing—with seething, roiling life, and life's spirits, being released in every moment of every day and every night upon this land.

I have thought often that the shape of the larch's body is like that of a candle flame. Broad at the base, measuring three, four, sometimes even five feet around, it maintains that barrel thickness for what seems like the entire rung of its length, before tapering quickly to a tip not unlike the sharpened end of a pencil.

This phenomenon is even more pronounced when the tapered tip gets knocked off by wind or lightning or an ice storm, leaving behind what now seems almost a perfect cylinder, and which continues living, even thriving, without its crown—able somehow to continue photosynthesizing and maintaining its vast bulk by the work of the few spindly branches that remain. Sometimes only a couple of such branches survive to nurture that entire pillar, so that one is reminded of the tiny arm stubs of another primitive, *Tyrannosaurus rex.*

There's some deal the larch have cut with the world, some intricate bargain, part vainglorious gamble and part good old-fashioned ecological common sense. They've cast their lot with the sun rather than the shade, having evolved to colonize new open space, such as that which follows a severe fire, or patches of forest that are infiltrated by slashes of light whenever other large trees fall over. Because of this, they race the other sun-loving trees—the pines and, to a lesser extent, the Douglas firs—for that position at the canopy where they can drink in all of the sun, where they have to suffer no one's shade.

But if they expend too much energy in that race for the sun—if they channel almost all of their nutrients into the vertical component of height at the expense of the horizontal component, girth—then they'll run the risk of being too skinny, too limber, and will be prone, then, to tipping over in the wind, or snapping under a load of ice or snow, or burning up like a matchstick in the first little fire that passes through; and what good is it then to gain the canopy—to win the race for that coveted position aloft—if only to collapse, scant years later, under the folly, the improvident briskness, of one's success?

When the larch and lodgepole are found together, as they often are up here, the larch will have been hanging just behind and beneath the lodgepole for those first many years, "choosing" to spend just a little more capital on producing thicker bark, both for greater individual strength—greater static strength—as well as to get a jump on the defense against the coming fires. It's always a question of when, not if.

However, as the lodgepole begin to reach maturity and then senescence, the larch begin to make their move; and as the lodgepole complete the living phase of their earthly cycle and begin blowing over, leaving the larch standing alone now, the wisdom or prudence of the larch becomes evident even to our often unobservant eyes. It is then that the true glory of the larch is manifested.

Seventy or ninety or a hundred years old by this point, the larch will have developed a thick enough bark, particularly down around the first four or five feet above ground level, to withstand many if not most fires.

And now, with the competition for moisture and nutrients removed, and the canopy more fully their own, the larch are free to really go wild. They didn't have to outcompete the lodgepole for those first seventy or a hundred years; they just had to tag right along behind and below. But now they can "release," as the foresters call it: having the canopy to themselves, they continue to grow slightly taller, but now pour more and more energy into girth, and into a thickening of their bark—battening down the ecological hatch against all but the most freakish, outrageous fires.

(So deep become the canyons and crevices, the corrugations, of that thickened bark, that a species of bird, the brown creeper, has been able to occupy and exploit that specific habitat: creeping up and down those vertical gullies, those crenulated folds a few inches deep, picking and pecking and probing for the little insects that hide beneath the detritus that collects in those canyons, and even building its nest in those miniature hanging gardens.)

Again, fire and rot are equal partners in this marriage, in this landscape quite unlike any other. And burn or rot, it makes no difference to the larch, really, how the lodgepole dies, for in their close association, the larch is going to feast upon the carcasses of the lodgepole and assimilate those nutrients, either in the turbocharged dumping of rich ash following the fire that consumes the lodgepole but only singes the thick bark of the larch or in the slower, perhaps sweeter and steadier, release of those same nutrients from the fallen lodgepoles as they decompose. The forest will always burn again, but on occasion fire may not return until after that dead and fallen lodgepole has rotted all away, has been sucked back down into the soil and then taken back up into the flesh of the larch, the larch assuming those nutrients as if sucking them up through a straw—which, in effect, it does, through the miracles of xylem and phloem.

At this point, of course, it's off to the races for the larch. They just get bigger and bigger, in the manner of the rich getting richer. And they seem to put all of this almost ridiculous bounty—sometimes, literally, this windfall—into the production of girth; they pork out, becoming still more resistant to the perils of fire and ice and wind, so that now time is about the only thing that can conquer the giants, and even time's ax seems a dull gnawer against the great larches' astounding mass and solidity.

It could be said that the growth of the lodgepole represents reckless imprudence and a nearly unbreakable flexibility, while the larch is all solidity and moderation. And if the lodgepole symbolizes the dense connections of community, and the notion that when one is hurt or bent, all are hurt or bent, then the larch is inflexible and isolate, the loner, seemingly independent in the world and as rigid, in his or her great strength, as the lodgepole is limber —the larch standing firm and planted, almost ridiculously so, in even the strongest storms, while all around the rest of the forest is swaying and creaking.

(Sometimes the force will be so great upon the larch that it'll snap and burst rather than bend, and its top will go flying off, cartwheeling through the sky like a smaller tree itself, and afterward, the damaged larch will set about its healing, sending up a slender new spar or sucker in place of the old top, cautious but

determined, and unwilling to cede anything, not even unto death
—remaining standing for a century or longer, even after the life
force has finally drained out of it.)

Western larch weighs 46 pounds per cubic foot, dried, here in
the Yaak; it's the heaviest, densest wood in the forest. It's like a
cubic foot of stone, standing or fallen.

Often they're so heavy, so saturated with their uptake of nutrients, that on the big helicopter sales, in places so far back into the
mountains or on slopes so steep that not even the timber industry's pawns in Congress will have been able to appropriate public
finances to build roads into those places, the sawyers will have to
girdle the big larch trees a year or more in advance of the logging.
This allows the life, the sap, to drain slowly out of the behemoths
so that when the helicopters do come, and the girdled trees are
finally felled, their dead or dying weight is considerably less than
if they were still green, and the helicopter companies are able to
save money on fuel because there's less strain on their engines.

As powerful and unyielding as the larch is, it only becomes
more so as it ages. You can read the individual stories, year by
year, in the growth rings sampled by an increment borer or, in the
case of one of the giants being felled, in the cross section made by
the saw and sawyer. The spaces between the growth rings expand
and contract through the years, charting the individual's explosive
early growth, and then the slowing-down, as if for a breath of air,
and then, when a fire or wind comes through and cleans out some
competitors, an expansion again, and to me such tales of thinning and thickening read like the scan of a kind of silent music—a
symphony of rise and fall, contraction and expansion, segue and
chorus.

Fire, ice, and wind: the larches' responses to and shapings by
these elements are dramatic, as is the flamelike alacrity with which
they leap from dormancy each spring, and with which they retire
for winter's slumber each fall—but I have to say, I think it is their
patience by which I am most impressed, and of which I am most
envious.

I want to believe they will be well suited to the coming temperature variations, the dormancy demanded not just by winter's extremes but by the coming heat and drought of global warming. I
know they lack the pines' flexibility, that they do not know how to
sway. Still, I believe in them, admire them, am in love with them.

And I dream of someone, one day, being able to walk from the summit of the Yaak to the Canadian border in a swath of un-interrupted old-growth larch ten miles wide, as once existed, as evidenced by the remnants still present, both standing as well as stumps. I dream that someday a hundred years from now a trav-eler—man, woman, child, or moose; bear, elk, wolf, or caribou—could set out on a warm summer's day and pass through the leafy cool light of an old larch forest, the duff soft underfoot and the air smoky and gauzy with the sun-warmed esters and terpenes ema-nating from the bark, and the odor of lupine sweet and dense all throughout the grove—and that the traveler could walk and walk and never leave the old growth; could walk all day and then into the night, through columns of moonbeam strafing down through the canopy, and still be within the old forest; could pass out of this country and into the next and still be among those old trees.

The shape and nature and spirit of this land would accommo-date such a vision yet. It is only up to our hearts to ask it.

I love the odor of them. I love the sight and touch of them. I love to lean in against them, to spread my arms around them, to touch the thick laminae of bark, to sit beneath them in storms while all else sways, as branches and streamers of moss whirl through the air.

I love to listen to the pileated woodpeckers drumming on them, and to the scrabble of little clawed animals scrambling up and down the bark of the living, as well as upon the fallen husks of the dead.

I love to see them lying on their sides in the ferns, rotting slowly—resting again, with the rain and sunlight still feeding somehow their magnificent and rotting bodies, even as they continue feed-ing the forest around them.

In their yearly dormancy as well, while losing the gold fire of their needles in autumn, they give back to the soil, particularly if a fire has just passed through, for the myriad wind-tossed casting of their needles acts as a net and helps secure the new bed of ash below, which might otherwise wash downslope and into the creeks and rivers, scouring the watercourse and eroding the soil.

It is a beautiful thing to see in the autumn, after a fire, those gold needles cast down by the millions upon a blackened ground. The two colors, black and gold, seem as balanced and beautiful as gold stars within the darkest night.

Late October and early November, after they have just gone to sleep, is the time I think of as being most their season. The sky above feels fuller in the absence of their needles. There is suddenly more space above, in a time when our spirits need that—in the dwindling days of light, and with winter's fog and rain and snow creeping in.

One night a damp wind blows hard from the south. In the morning the hills and mountains are covered with gold. It's an incredible banquet, a visual feast, and our eyes take it in all at once, and a thing stirs in our blood, a strengthening and quieting-down both; and farther back in the forest the bears begin to crawl into their dens, seeking sleep also.

If the gold needles had stayed up there against that cerulean October sky forever, surely we would have eventually gotten used to them, and taken them for granted.

Hiking down off a mountain from far in the backcountry, I stop at dusk, weary, and without shedding my burdened pack take a seat on an old fallen larch, one of those ancient giants from the last century, its heartwood finally rotting but its outer husk still firm.

The immense log is covered completely with the gold confetti of its descendants growing all around it, and there is no table or other furniture I have ever seen more elegant or beautiful than that impromptu bench, nor more timely—I was tired and needed a place to rest, so I sat down and it was there for me—and I sit there resting for a long time, watching the dusk give itself over to dark.

And just as there is no furniture that could be the equal of a fallen larch left in the woods to rot or burn at its own pace, or under the pace of this landscape that is so intensely its partner, surely there can be no gold-lined streets of heaven superior to what awaits the residents of this valley on a fine October morning after a night during which the wind has blown hard, when our dreams of a night sky filled with swirling, shimmering gold are exceeded only by the beauty of reality as we first step outside to see one more glorious season being born into the ceaseless and enduring world.

BRETT FORREST

Shattered Genius

FROM *Playboy*

I HAD NEVER BEEN on a stakeout, but I knew how it was done. I took a book. I brought a few sandwiches. I flipped on the radio and listened to the traffic report in Russian. That kept me awake as I waited for the mathematician.

I'd first heard of Grigori Yakovlevich Perelman about nine years ago, as news of his achievement leaked beyond the international mathematics community into popular headlines. Word was that someone had solved an unsolvable math problem. The Poincaré Conjecture concerns three-dimensional spheres, and it has broad implications for spatial relations and quantum physics, even helping to explain the shape of the universe. For nearly one hundred years the conjecture had confused the sharpest minds in math, many of whom claimed to have proven it, only to have their work discarded upon scrutiny. The problem had broken spirits, wasted lives. By the time Perelman defeated the conjecture, after many years of concentrated exertion, the Poincaré had affected him so profoundly that he appeared broken too.

Perelman, now forty-six, had a certain flair. When he completed his proof, over a number of months in 2002 and 2003, he did not publish his findings in a peer-reviewed journal, as protocol would suggest. Nor did he vet his conclusions with the mathematicians he knew in Russia, Europe, and the United States. He simply posted his solution online in three parts—the first was named "The Entropy Formula for the Ricci Flow and Its Geometric Applications"—and then e-mailed an abstract to several former associates, many of whom he had not contacted in nearly a decade.

I liked his style. The more he did, the more I liked. In 2006 Perelman became the first person to turn down the Fields Medal, the top award in mathematics (there is no Nobel Prize in math). He has declined professorships at Princeton, Berkeley, and Columbia. In 2010, when the Clay Mathematics Institute in Cambridge, Massachusetts, awarded him a $1 million prize for proving the Poincaré Conjecture, Perelman refused it. Unemployed these past seven years, he lives with his mother in a former communal apartment in St. Petersburg, the two subsisting on her monthly pension of $160. "I have all that I need," Perelman has told his concerned Russian math colleagues, with whom he has severed all but the most perfunctory telephone relations.

Perelman last gave an interview six years ago, shortly after a collective of PhDs finished a three-year confirmation of his proof. Since then, the domestic and international press have harassed him into reclusion. Perelman has spurned all media requests, muttering tersely through his apartment door against a wave of journalists. "I don't want to be on display like an animal in a zoo," he told one reporter. "My activity and my persona have no interest for society." When one journalist reached him by phone, Perelman told him, "You are disturbing me. I am picking mushrooms."

While Russian society has largely passed judgment on Perelman —misanthrope, wacko—I admired him for his renunciation of the modern world's expectations, his devotion to labor, his results. He had not solicited fame or reward in proving the Poincaré, so why should he be required to react to public notice? His will was free, his results pure, and therein lay his glory.

There was more than one path to glory, I reasoned, and some glory might be found were I to solve this riddle. Perelman was the riddle, speaking through mathematics, the complex language of his Poincaré proof incomprehensible to all but a few hundred mathematicians. For the rest of us, eager to grasp the meaning of exceptional behavior, there was only silence. With slight hope, I booked my ticket to St. Petersburg.

In advance of my trip I phoned Sergei Kislyakov, the director of St. Petersburg's Steklov Institute of Mathematics, where Perelman had worked as a researcher. In late 2005, two years after his Poincaré proof had made him the biggest name in his field, Perelman handed Kislyakov his resignation, stating that he had been "disappointed" in math. He was abandoning math altogether, he said.

Kislyakov knew how obstinate Perelman could be. When I explained that I planned to speak with Perelman, Kislyakov interrupted me. "I discourage you from coming here," he said. "Perelman talks to no one, but he particularly hates journalists."

"My editor has told me to go," I explained.

Kislyakov sighed. "Then I guess you must."

It was spring. St. Petersburg was preparing for the Victory Day parade. Tanks lined the central canals. Banners crested the streets. In Kupchino, the southernmost stop on the blue Metro line, far from the palaces that give Petersburgers their proud self-possession, it looked like any other new day. The red-and-white trolleys coursed up the grassy center lanes of the avenues. People strolled in the courtyards that connected the battered housing projects. Russian prime minister Dmitry Medvedev had grown up in Kupchino, but this neighborhood was so removed from fame and influence that it made a perfect home for someone who preferred to escape all notice.

In my search for Perelman, I thought I might rent an apartment, find one with a good view of his building's entrance. A real estate agent walked me all over the neighborhood. "Isn't there a well-known scientist around here?" I ventured casually.

"He lives somewhere on this street," said the broker.

"Have you ever seen him?"

"Seen him?" he said with a laugh. "Sure, I've seen him. Like I've seen Putin—on TV." The guy showed me one dump after another.

To get around, I rented a Hyundai, all that was available at the leasing agency downtown. I parked outside Perelman's building. A dozen stories high, made of unadorned concrete panels in the dull Brezhnev style, the structure covered half the block. A handful of people gathered in front of the brown steel door to Perelman's stairwell, smoking, passing around a morning beer. In this place it appeared there was little rush to achievement.

On a previous day I had met one of Perelman's neighbors, a teacher at a local school. She said that she and others in their building joked about pleading with Perelman to accept the $1 million prize on their behalf. I couldn't tell which was the source of greater amusement to her, the idea that Perelman would accept the million or the idea that he would engage her in conversation.

Perelman mixed with no one, she said, refusing even to ride the elevator unless he was the only one in it.

And with whom would he mix? The people I saw were roughly drawn, the elderly leaning on spindly wooden canes, the teenagers darting between the kiosks, wasting the day. An androgynous bum with dirty blond hair nosed around the garbage. An old lady in a coarse gown looked at me through the windshield, then spat.

Ragged as these surroundings were, Perelman exceeded them. As a younger man, he had been handsome, with soft, dark features. But recent pictures—taken with a cell-phone camera in a subway car and then transmitted across the web—projected a different image. Perelman's clothes were dirty and rumpled, his black beard mangy. Ringing the bald crown of his head was a nest of hair that stood on end. He looked disturbed as he gazed out from under thick eyebrows, chewing a nail. How would he react when I approached him?

My mark did not appear that first day, and I cautioned myself to have the patience of Perelman. He had spent seven years proving the Poincaré Conjecture, seven years displaying the sort of patience that is well beyond most people. The editors of one Russian tabloid ran out of patience tracking him. When they sent a reporter to Kupchino, the reporter got nothing. A female clerk said she had once exchanged a few words with Perelman. The next morning the headline read, THE SECRET LOVE OF GRIGORI PERELMAN.

When I met Sergei Rukshin, Perelman's closest friend, I realized that my respected counterparts in the Russian press had complicated my task. "Nice to meet you," I said when I arrived at Rukshin's office in a St. Petersburg high school. He replied, "We'll see if it will be nice or not." But like a rusty faucet, once turned, Rukshin gushed, speaking about Perelman for more than four hours.

It was Rukshin, serving as the instructor of a specialized Leningrad math club, who recognized Perelman's talent in 1976. It was Rukshin, along with other supporters in academe, who piloted Perelman through the anti-Semitic Soviet policies that nearly prevented the young Jewish genius from obtaining an education commensurate with his mind. And it is Rukshin who now grieves over the condition of this favored pupil: "He lives in a blockade."

*

Day two of my stakeout. A truck pulled up and parked, obstructing my view of the entrance to Perelman's wing of the building. As I opened the door of my car, a few guys with fresh cuts on their faces straggled by carrying a ten A.M. bottle, looking for something to do. I stayed where I was, grazing on chips, and kept my eyes on either side of the truck, where I could still see people passing by. A man in an ink-black coat appeared in front of my car. He waved his hands at me wildly, yelling, "No, no." Then he turned away. I couldn't figure out what that meant, except to say that the locals were beginning to notice me. The potential for violence mounted hourly.

There wasn't much I could do about that, and I thought instead about Perelman's evolution. Rukshin told me that as a child Perelman had interacted with other students, that he had not been antisocial. Besides math, he enjoyed Ping-Pong and the opera. According to Rukshin and others who have known him since adolescence, Perelman is heterosexual, but as Rukshin noted, "If Grisha ever looked on anything with loving eyes, it was on the blackboard." No friend can recall the name of a girlfriend. Shortly after Perelman earned his PhD, the Soviet Union collapsed. He left for the United States, where he performed postdoctoral research at NYU, Berkeley, and the State University of New York at Stony Brook. He was out in the world, interacting with contemporaries. He was doing things.

Yet he was already turning inward. When the top mathematicians in Russia were earning roughly $100 a month in salary, Perelman was exposed to a Western world of tenured professors, academic grants, and funded research labs—the business side of academe. "It's possible to sell a theorem and it's possible to buy it," he told Rukshin when he returned to Russia, disenchanted, in 1995. "Even if you don't have anything to do with it."

Perelman had already begun his work on the Poincaré Conjecture, a theorem expounded in 1904 by Henri Poincaré, a French polymath and the founder of topology, the mathematical study of abstract shape. Because the problem had a history of false proofs, Perelman told no one about his work lest he be discouraged. He was also wary that unsolicited input would cloud his mind. "For Grisha, it was complete self-restriction," Nikolai Mnev, a friend and former colleague of Perelman's, told me.

Had I such industry, my life might have carried me to a position loftier than the seat of a Hyundai on St. Petersburg's provincial fringe, waiting for someone who would be displeased to see me, should he appear at all. The hours passed. I bit into a sandwich, bundled my Windbreaker and used it as a pillow.

Who was I to complain? Perelman had truly suffered, acutely. He withstood a claim—since refuted—on his Poincaré proof from a rival Chinese mathematician. He turned down the Fields Medal, believing that acceptance would be, as Rukshin explained, fundamentally dishonest. Perelman once rebuffed a TV crew from Russia's Channel One when they barged through his apartment door, pushing aside his mother. He withstood the procrastination of the Clay Mathematics Institute, which took its sweet time—five years—to offer him the $1 million it had committed to the person who solved the Poincaré. "Grisha is tortured by the imperfection of humanity," Rukshin said.

All this was going through my mind when, suddenly, Perelman appeared. Over a field of parked cars, his wild hair bounced along as he walked away from me on the path by his door. I had to chase after him. I opened the car door. When I looked up from the handle, relocating my mark, I saw that it was not Perelman. It was simply a man with wild hair, fleeing the pleas of the androgynous bum.

It was day three of the stakeout, and still no sign of Perelman. I was secretly relieved, since I had no idea what to ask him. I'm not much of an interviewer. I approach my subjects as if we were in a bar, chatting over a beer. A standard swindle but enjoyable, if I'm not mistaken. People like talking about themselves. You just have to give them the chance.

But how do you talk to somebody who doesn't talk to anybody? Every question I thought to ask, I knew Perelman wouldn't answer. I couldn't take direction from the Russian press, which had deluged him with questions about why he wouldn't accept the money, why he had turned down the Fields Medal, why he wouldn't talk to them.

I didn't want to bother Perelman. I didn't want to be like all the others who had forced him into exile. I believed there was a delicate way to approach him.

I consulted those who knew him. When I met with Alexander Abramov in Moscow, he described the last phone call he had had with Perelman, three years prior. Abramov, a professor, has known Perelman since 1982, when he coached the Soviet team at the International Math Olympiad. (Perelman won a gold medal, posting a perfect score.) Exasperated by Perelman's solitude, Abramov asked him what he should do in order to meet with him. Perelman suggested that Abramov move to St. Petersburg. "Forever?" Abramov asked. "Maybe," said Perelman before hanging up the phone.

Maybe Perelman didn't like Abramov anymore. Maybe he didn't like anybody anymore. "I'm afraid he is at the level of a nervous breakdown," Rukshin said. "If this was still the Soviet Union, he would be forced into psychiatric treatment for this behavior." In 2008 Perelman asked Rukshin to limit their phone calls. Now they speak about once a year.

"It looks very much like the story of Bobby Fischer," Abramov said. "And Bobby Fischer couldn't be called a happy man."

It was the afternoon of day three, and the androgynous bum pleaded through my car window for a few rubles. Even up close I could not tell if this was a man or a woman. I watched the bum move along a little richer. When I refocused my eyes on Perelman's door, I heard myself gasp, "There he is!"

It was Perelman all right. The beard, the hair, the expression of uncertainty as he stumbled into the sun with his mother, Lyubov, by his side. He shuffled toward the garbage bins stacked by the door, looking as if he might rummage through them. He wore a black ski jacket, a black shirt, black pants. His mother was dressed in a red overcoat and a white beret. They turned down the lane, heading toward the courtyard behind their building. I locked the car.

The courtyard was the size of a city block, with trees, parking lots, and playgrounds. Trailing at a considerable distance, I saw Perelman and his mother moving across a grassy field. I decided to approach him head-on rather than sneak up from behind, taking all measures to avoid agitating him. Even though I knew he had known English quite well at one time in his life, I thought it best to speak Russian with him to put him at ease.

I walked along one edge of the courtyard, hoping to meet him as he reached its far side. I hurried past a trash heap, around the fencing of a dead tennis court. I circled around a small school, and when I reached the far edge of the grassy field, Perelman and his mother weren't there. I had lost them.

Frantically I searched the courtyard. I located them again, along a row of parked cars. But when I made another loop in order to get in front of them, I didn't see them. When I spotted Perelman and his mother once more, they were heading back the way they had come. I didn't have the luxury of positioning. I would have to approach them from behind. I walked briskly. I was 20 yards from Perelman and closing. Still I didn't know what to say.

Then I was at his side, and there was no more time to think. "Grigori Yakovlevich?" I said, employing his middle name, in polite Russian form. "Is it you?"

Perelman's head rotated slowly. He appraised me from the corner of one eye. He said nothing. "Excuse me, please," I continued. "I don't want to bother you. But I have come from America to speak with you."

Up close, Perelman looked about five-foot-ten and slighter than I had imagined. He was less menacing than he appeared in pictures. He did not waste thought on his appearance, though. Dandruff caked the shoulders of his coat. His clothes were streaked with stains.

Perelman spoke with a high-toned, birdlike voice. And he knew what to say. "You're a journalist?" he asked. His mother peeked at me from behind his shoulder, then pulled away. I nodded. Perelman looked at the sky, letting out a pained sigh. We took several small steps together. "From which publication?" he asked.

I told him. He nodded in recognition but said, "I don't give interviews."

"I know," I said. "That's okay." Perelman and his mother stopped walking. They looked me up and down, as though what I'd said had confused them. I didn't know how this was going to go, but at least Perelman had not run away. So I put on a big smile. "Good weather today, huh?" I said. And to my surprise, both the terrifying recluse and his nervous mother let out a laugh. They were disarmed. I was in.

*

"How did you know we would be here?" Lyubov Perelman asked, stepping out from behind her son. She wore thick glasses, and her cheery face puffed out beneath the beret.

"I'm embarrassed to say," I told her.

"Well?" she said.

I nodded toward the street. "I've been sitting in a car out there waiting for you."

"Really?" she said.

"It wasn't so bad," I said. "I had a book."

"How did you find out the address?" Perelman asked me.

"I have a connection," I said. "With the police."

His eyes went wide. "The police?" he said. "Are you Russian?"

"American."

He looked at me curiously. "Are you sure you're not Russian?" By all signs that I could interpret, Perelman was eager to speak with me, glad for human contact.

"Do you mind if I walk with you for a little bit?" I asked. Perelman shrugged, and we kept on. He had laughed once, I thought. Maybe he would laugh again. "I was nervous," I told him. "Everybody says you are frightening." Perelman squinted at the sky as if contemplating something I would never understand. A man passed in front of us, walking a cat on a leash.

Lyubov Perelman said, "If you're not getting an interview, what's the point of this?"

Perelman put his arm around her. "It's okay, Mother," he reassured her. "We're just walking."

Considering all I had learned about Perelman, this display of considerate behavior amazed me. And it emboldened me. No one had gotten this close to him in years. "I understand you're not practicing math anymore," I said. "Can you tell me what you are working on?"

"I have left mathematics," he said. "And what I'm doing now, I won't tell you."

I was ready with another question, but he had one of his own. "You're really not Russian?" he asked. "You speak like someone who was born in Russia and left at eight or nine, then came back as an adult. You have this sound."

Pressing my momentum, I asked him a few easy questions, hoping to open him up. "What are your plans for the May holidays?"

"Did you enjoy your time in America?" "How often do you take these walks?" Each time, Perelman shrugged, stared into the sky, and said nothing. I wasn't sure if he had heard me. I looked at his mother, and she raised her eyebrows as though she didn't know what to say either. A smile crossed her face.

We made our way toward the archway that led to his entry. I tried another serious question. "Considering your abilities and how young you still are, how might you return to science?" He wheezed. After a short silence, his mother asked if I was wearing a wire.

I resolved to draw him out once more. Trying to build common ground, I touched on the similarities between writing and mathematics, emphasizing the solitude that each discipline required. I looked at him with an open, friendly face. He stared again at the sky, a blank page.

We reached the archway and stopped. Perelman and his mother stared at me, wondering how this would end. I looked at Perelman and asked, "How's your Ping-Pong game?"

"I haven't played in a long time," he said. He laid an arm across his mother's shoulders. He was becoming agitated. We had walked and talked for twenty minutes, and what had I figured out? I had gotten a feeling for the man, but I had not solved the riddle. Would he help me do it? There was time for one final question. I put it to him in English, the single philosophical question that I hoped he would consider. "Where does your life go from here?" I said.

Perelman stepped closer to me. I saw that one of his upper teeth was dark brown, decayed. "What?" he said, his English skills apparently dormant. Perelman's face was focused in concentration as I repeated the question, and I thought that he might answer it. But when I finished speaking, his face went slack, as before. He understood what I wanted to find out, the path of this unusual life. He mumbled, "I don't know."

We said our good-byes.

Walking to my car, I felt as though I had failed, having gained such rare proximity to Perelman, only to have the man slip through my grasp. But then I paused, for there must have been something I had missed. Perelman was as unadorned yet just as complex as the conjecture he had proven. He had relieved the Poincaré Conjecture of its mystery, and in so doing had replaced it, becoming the puzzle himself, granting the world knowledge, yet diminishing its enchantment not at all. We don't have to figure out everything. The unknown has its own value.

Through the windshield of the Hyundai, I watched Perelman and his mother approach their entryway, the bums and the kids and the new mothers of Kupchino going about their lives. Perelman and his mother retreated into the darkness of the vestibule. The metal door slammed closed behind them. Perelman was out, he was in. He had gotten some air.

JEROME GROOPMAN

The T-Cell Army

FROM *The New Yorker*

IN THE SUMMER of 1890, an adventurous seventeen-year-old
from New Jersey named Elizabeth Dashiell traveled across the
United States by train. During the journey, she caught her hand
between the seats of a Pullman car. The hand became swollen and
painful, and when it didn't heal after she returned home, Dashiell
consulted William Coley, a young surgeon in New York City. Unable to determine a diagnosis, he made a small incision below the
bottom joint of her pinkie finger, where it connected to the back
of her hand, to relieve the pressure, but only a few drops of pus
drained out. During the following weeks, Coley saw Dashiell regularly. In the operating room, he scraped hard, gristly material off
the bones of her hand. But the procedure gave only fleeting relief.
Finally, Coley performed a biopsy that showed that Dashiell had
sarcoma, a cancer of the connective tissue, which was unrelated
to her initial injury. In a desperate attempt to stop the cancer's
spread, Coley followed the practice of the time and amputated
Dashiell's arm just below the elbow. But the sarcoma soon reap-
peared, as large masses in her neck and abdomen. In January
1891, she died at home, with Coley at her bedside.

After Dashiell's death, Coley was distraught and searched
through the records of New York Hospital for similar cases. He
found one patient who stood out from the grim stories. Eleven
years earlier, Fred Stein, a German immigrant who worked as a
housepainter, had a rapidly growing sarcoma in his neck. After
four operations and four recurrences of the cancer, a senior sur-

geon declared Stein's case "absolutely hopeless." Then an infection caused by streptococcal bacteria broke out in red patches across Stein's neck and face. There were no antibiotics at the time, so his immune system was left to fight off the infection unaided. Remarkably, as his white blood cells combated the bacteria, the sarcoma shrank into a bland scar. Stein left the hospital with no infection and no discernible cancer. Coley concluded that something in Stein's own body had shrunk the cancer.

Coley spent the next decade hoping to replicate Stein's extraordinary recovery. In *A Commotion in the Blood,* published in 1997, Stephen S. Hall describes how Coley inoculated cancer patients, first with extracts of streptococcal abscesses, termed "laudable pus," and later with purer cultures of the microbes. He claimed several successes, but the medical establishment did not embrace his approach, because his results could not be reliably reproduced. His primary critic, the pathologist James Ewing, believed that the new technique of radiation was the only scientifically sound way to treat cancer.

Coley's work was financially supported by John D. Rockefeller Jr., a classmate of Dashiell's brother who had considered Elizabeth his "adopted sister." But Rockefeller also donated to Ewing's research. While Coley told stories of miraculous recoveries, Ewing presented numbers that consistently demonstrated the power of radiation. Ultimately, Rockefeller chose Ewing as his scientific adviser. Rockefeller's support led to the creation of what is now the Memorial Sloan-Kettering Cancer Center, one of the foremost institutions studying and treating malignancies. The idea that the body's immune system could play a crucial role in eradicating cancer was largely discarded. One doctor at the time called Coley's hypothesis "whispers of nature."

In the last hundred years, progress in the treatment of cancer has come mostly from radiation and chemotherapy. Previously fatal blood-cell cancers, such as childhood leukemia and Hodgkin's disease, are now curable. But solid tumors, which grow in the lungs, the colon, and the breast, have stubbornly resisted treatment once they spread beyond their initial site.

In 1971 the Nixon administration declared a "war on cancer," promising Americans that within ten years the disease would be beaten. At the time, many researchers believed that cancer was caused by a virus that speeded up a cell's metabolism, resulting in uncontrollable growth. After all, they had discovered some hundred viruses that caused cancer in amphibians, birds, and mammals. In the early seventies, interferon, a drug that had been developed from a protein released by white blood cells during a viral infection, was widely thought to be a possible cure for cancer; in 1980, it appeared on the cover of *Time.* The tumors of mice shrank dramatically when treated with the drug. But in patients interferon failed to cure solid tumors, and melanoma responded only occasionally.

Over the next decade, other proteins produced by the body as part of its immune response were made into drugs, most notably one called interleukin-2. In 1988 Armand Hammer, the ninety-year-old oil-company magnate who chaired Ronald Reagan's cancer panel, sought to raise a billion dollars, with the aim of curing cancer by his hundredth birthday. He touted interleukin-2 as an immune booster that could achieve the goal. But most solid tumors were impervious to it, too.

In the past fifteen years, as tumors have been found to contain genetic mutations that cause them to grow unrestrained, the focus of research has shifted to cancer's genome. Targeted therapies designed to disarm these mutations are now at the forefront of care. The first successful targeted therapy was Gleevec, which caused rapid remissions in chronic myelogenous leukemia, with few and mild side effects. Herceptin, a targeted therapy that attacks HER-2, a protein that is found in some 20 to 30 percent of breast-cancer cases, has also been effective.

Advances such as these caused Coley's approach to fade into obscurity. Harold Varmus, a Nobel laureate and the director of the National Cancer Institute, told me that until very recently, "except for monoclonal antibodies, every therapy that exploited the immune system was pretty abysmal. There weren't any good ideas about why immune therapy failed." But now patients who did not respond to available therapies have shown dramatic and unexpected responses to a new series of treatments that unleash the immune system. Coley's theories are suddenly the basis for the most promising directions in cancer research. In March 2011, the National Cancer Institute announced that it would fund a network of twenty-seven universities and cancer centers across North America to conduct trials of immune therapies. Mac Cheever, the director of the program, who is at the Fred Hutchinson Cancer Research Center in Seattle, described it as a way to speed the practical work of developing treatments. "All of the components needed for effective immunotherapy have been invented," he said.

Jim Allison, the director of the tumor-immunology program at Memorial Sloan-Kettering, began his career as a researcher at the University of Texas Cancer Center in 1978. At the time, he was taken with the idea that the T cell could be directed against cancers. T cells, a potent type of white blood cell, destroy cells infected with microbes that they recognize as foreign. The immune system uses a variety of white blood cells to fight disease. Some, like neutrophils and macrophages, engulf and chew up microbes. In contrast, T cells attack the microbe from the outside, with a fusillade of enzymes. Cancers disarm the immune system, producing proteins that cause T cells to either quickly become exhausted and die or blithely overlook the tumor. Allison's research focused on why T cells failed either to recognize cancer as being aberrant or to attack it, as they do with microbes.

Allison's mentors discouraged him from pursuing research on T cells. "Tumor immunology had such a bad reputation," he told me when we met in December at his laboratory at Sloan-Kettering, which overlooks the East River. Allison, who is sixty-three years old, is a thickset man with a stubbly beard and a gravel voice. "Many people thought that the immune system didn't play any role in cancer." Treatments like interferon and interleukin-2 had led scientists on a roller coaster of hype followed by disappointment. Immune therapy was also tainted by popular claims that following a certain diet or reordering your mind could be natural immune-boosting ways to cause tumors to disappear, with none of the miserable side effects of chemotherapy and radiation.

But Allison started looking at how the immune system fights disease, using mice as study models, and capitalized on a critical discovery: T cells require two signals to attack a target effectively. The first signal, he said, was "like the ignition switch," and the second "like the gas pedal." When working against a microbe, both signals were operative. But in the presence of cancer, "T cells don't get those signals to attack," he explained. Allison started to wonder what it would take to reliably activate the immune system against cancer.

In 1987 researchers in France discovered a protein called cytotoxic T-lymphocyte antigen-4, or CTLA-4, which protruded from the T cell's surface. "There was a real race among a number of labs to figure out its function," Allison recalled. A scientist at Bristol-Myers Squibb, using results from his lab, contended that CTLA-4 increased the activity of T cells and the immune system. But Allison and Jeffrey Bluestone, an immunologist, obtained results from independent experiments that contradicted that conclusion. Allison and Bluestone believed that CTLA-4 actually acted as a brake on the T cells, and Allison thought that it might be keeping the immune system from attacking tumors. "Jeff and I were kind of in the wilderness for a while," Allison said. "Before this, people just thought that T cells died on their own." He speculated that treatments designed to activate the immune system might have failed because the treatments were actually stimulating CTLA-4. As Allison put it, "We ought to free the immune system, so it can attack tumor cells."

Allison's postdoctoral researchers implanted cancer cells under the skin of mice, some of which were then treated with an antibody that blocked CTLA-4. After several weeks, the cancers disappeared. One of the researchers showed Allison the data in early December 1995. Allison was astounded. The lab was about to go on Christmas break, but he wanted to repeat the experiment immediately. "I told the researcher that he should inject the tumors into a new group of mice, and have a control group that didn't get the antibody. And I'd measure the tumors myself," Allison recalled. "So it was really a blinded experiment, because I didn't know what was what." A week later, Allison measured the cancers. "The tumors were still growing, and I'm starting to despair. And then, in half of the mice, the tumors just seemed to stop, but in the other half of the group they kept going. And then in the ones in which it stopped, the cancer started disappearing and just went away." Allison added, "It immediately confirmed our original assumption that this could be good for any kind of cancer."

For two years, as Allison continued his experiments on mice, he approached pharmaceutical and biotech companies for help in developing the treatment for patients, but he was repeatedly turned away: "People were skeptical of immunology and immune therapy. They would say, 'Oh, anybody can treat cancer in mice.' Sometimes they'd say, 'You think you can treat cancer by just removing this negative signal on a T cell?'"

Allison also learned that Bristol-Myers Squibb had filed for a patent asserting that CTLA-4 stimulated T-cell growth. "If that was the case, you would never, ever think about injecting an antibody that blocked CTLA-4 into a cancer patient, because it would make things worse," he said. "People were scared of putting that into a patient." But Allison persisted, telling industry executives that Bristol-Myers Squibb was wrong. Finally, he persuaded a small company called Medarex to invest in the approach.

Among its first trials on humans, in 2001, Medarex included patients with malignant melanoma, because it was one of the few cancers that had occasionally responded to immune-based treatments like interferon or interleukin-2. In pilot studies, patients were treated with the antibody to CTLA-4, and, as in mice, the cancers continued to grow for some weeks before a few of the tumors shrank. In 2004 Bristol-Myers Squibb formed a partnership with Medarex to collaborate on the drug. A subsequent trial showed scant impact after twelve weeks. Many of the tumors got bigger, and in some patients new lesions appeared. Pfizer was also testing an antibody to CTLA-4 and concluded that it was a failure; the trial was stopped early.

Months after the end of the Bristol-Myers Squibb study, however, several of the clinicians involved, including Jedd Wolchok, of Memorial Sloan-Kettering, and Stephen Hodi, of the Dana-Farber Cancer Institute in Boston, realized that the tumors had either stopped growing or begun to shrink. Wolchok and his colleagues prevailed upon Bristol-Myers Squibb to include overall survival rates of patients after several years. (Because the established criteria for judging the effectiveness of chemotherapy drugs are based on the first months of treatment, the trial had been considered a failure.) "It was pretty courageous," Allison said, "because it would take a long time to finish the study." In June 2010, the results were presented at the annual meeting of the American Society of Clinical Oncology. Although the drug had extended the patients' lives a median of only four months, nearly a quarter of the patients were alive two years into the trial. Their predicted survival had been seven months. "This is a drug unlike any other drug you know," Allison said. "You are not treating the cancer—you are treating the immune system. And it was the first drug of any type to show a survival benefit in advanced-melanoma patients in a randomized trial."

Allison's results astounded cancer specialists. *Nature* published a review in December 2011 and noted that the antibody to CTLA-4 "provides realistic hope for melanoma patients, particularly those with late stage disease who otherwise had little chance of survival. More broadly, it provides clear clinical validation for cancer immunotherapy in general." I asked Harold Varmus why Allison had had success where other researchers in immunotherapy had failed. "We need to understand what we do," he said. "Jim made things understandable."

"You've got to be careful about using the word 'cured,' because some patients have residual tumors," Allison said. "But it doesn't matter, because their cancers are not growing. And, in others, tumors just pop up and then go away. So it's become something of a chronic condition," rather than a death sentence. Allison moved to Sloan-Kettering to be closer to the clinical trials conducted by Wolchok and others. "I just wanted to be the advocate who is keeping it in everybody's face," he said.

In the fall of 2003, Sharon Belvin was a twenty-two-year-old student teacher with plans to marry the following June. She ran between four and five miles a day and began to notice that her chest hurt after her morning workout. The student health service thought that she might have viral bronchitis, picked up from the children in her class. But her symptoms did not improve, and she was given other diagnoses, including asthma and pneumonia. Before long, she found it uncomfortable even to walk. On a visit to her mother, Belvin saw the family physician, who found a lump on her clavicle. A biopsy showed that she had metastatic melanoma. "It shocked me," Belvin told me. "I was never a sunbather. And I never had any lesions on my skin." A week before her wedding, she completed her evaluation. A body scan "lit up like a Christmas tree," she recalled. "I ended up having chemotherapy on Monday, Tuesday, and Wednesday, and got married on Saturday." During four months of therapy, the tumors shrank a bit. Then they began to grow again. An MRI showed that the melanoma had spread to her brain. Belvin went to Sloan-Kettering, where the brain tumor was treated with radiation. After recovering from the procedure, she received interleukin-2, to stimulate her T cells. The therapy caused such a severe reaction that "my skin peeled off all over my body," Belvin said. "I was so violently ill, I don't remember half of what happened." Worse yet, the treatment failed to stop the cancer's growth. "The doctor told me, 'If you are going to take a vacation, you'd better do it now.'" Belvin and her husband went on a Caribbean cruise.

When she got back, Belvin returned to the hospital and had twelve liters of fluid drained from her chest. Then Wolchok offered Belvin treatment with the antibody to CTLA-4, which was still an experimental therapy. "By that point, I had told my husband, 'If this doesn't work, I don't know how much more I can take,'" she recalled. Wolchok gave her an informed-consent release that listed all the possible side effects. "It was pages and pages of this could happen to you and that could happen to you. I didn't read one page. I just signed at the bottom and said, 'Give it to me.'"

The antibody was infused through one of Belvin's veins, and she had a drastic reaction: her body shook and she experienced drenching sweats, as well as an immune attack on her thyroid gland. "I thought I was dying, the rigors were so bad," she recalled. After four treatments given every three weeks, Belvin went for a set of scans. "I remember how Dr. Wolchok came in with this huge smile on his face, and he was like, 'This is great!' He was just floored." The massive tumors in her lungs had shrunk significantly.

Wolchok did not want to raise Belvin's hopes too much. But "every single scan that I had after that time, the tumors kept shrinking," she said. Eight years after her diagnosis, she still has no signs of the cancer.

Belvin's case is remarkable, but it contradicts the popular notion that boosting the immune system is a "natural" way to treat cancer, free of the harsh side effects associated with chemotherapy or radiation. The results of immunotherapy can include an attack on the skin, intestines, lungs, liver, thyroid, pituitary gland, kidneys, and pancreas. When T cells are stimulated to an intensity that destroys cancer cells, they can also cause collateral damage to normal tissue. Wolchok told me, "You may need to cross the line to toxicity for the immune system to be effective against a cancer.

It's not a free ride." Because Belvin's thyroid gland was destroyed by the therapy, she now requires replacement hormones.

Steven Rosenberg, the chief of surgery at the National Cancer Institute, who played a key role in developing interleukin-2, also conducted some of the early studies with the antibody to CTLA-4. He noted that the bowel often became severely inflamed with the treatment: "You have, like, eight liters of diarrhea a day. The colitis is atrocious and would be lethal in almost everybody. If you don't put those patients on corticosteroids immediately, they'll die."

"In the field of oncology, the bar is set so low," Rosenberg told me. He welcomes the outcomes for patients like Belvin but is cautious about the long-term benefits of similar treatments. "I believe that the antibody to CTLA-4 will cure some patients with melanoma, although the follow-up is short." But unless all detectable cancer disappears, he said, "the tumors are going to grow back eventually."

Rosenberg has pioneered a different strategy, called "adoptive cell transfer," in which T cells are taken from a patient's tumor and given immune stimulants such as interleukin-2, which cause them to replicate. Then they are put back into the body. In the latest of three trials of patients with melanoma who underwent adoptive cell transfer at the National Cancer Institute, nine of twenty-five patients have been in complete remission for more than five years. Across all three trials, five patients who had received earlier, unsuccessful treatment with the antibody to CTLA-4 are in remission.

Sam Breidenbach, who runs a construction company in Wisconsin, was one of those five. In September 1999, his wife noticed a small mole on his back. He went to the hospital at the University of Wisconsin in Madison and was told that he had melanoma. It was caught early, and the doctors, after removing it, said that the cancer did not appear to have spread. But three years later, while playing volleyball, he lunged to spike the ball, and felt a pull at his left flank. "It was this roly-poly little nodule on my left hip, at the top of the bone"—a metastasis from the original melanoma. "A local oncologist just basically said, 'You'll be lucky to live five years,'" Breidenbach recalled. He returned to the hospital in Madison, where he was given high doses of interferon. "For the first month, I was just totally dead. I couldn't do anything." The treatment was ineffective. Within months the melanoma had appeared in the lymph nodes of his left groin.

Breidenbach found out about Rosenberg through his daughter, who was in a violin class with a girl whose father had been treated for melanoma at the National Cancer Institute. Breidenbach contacted Rosenberg, who treated him with an experimental melanoma vaccine. Breidenbach did not respond to the treatment, and the melanoma spread to his liver and lungs. In the summer of 2003, after being treated with the antibody to CTLA-4, he developed excruciating pain in his abdomen—pancreatitis, caused by the toxicity of the immune response. "It was so brutal that they had to stop the treatment," Breidenbach said. "They were basically out of any other ammunition to throw at me." His doctor at the University of Wisconsin told him that he couldn't expect to live more than four to six months. One oncologist suggested chemotherapy, but "I knew the numbers, and my wife and I said, 'If this is really the remaining time I have on the planet, why make it miserable?'"

Over the week of Thanksgiving, Rosenberg called and told him that his research team had studied his T cells in the laboratory. "Your cells are jumping out of the petri dish," Rosenberg said. He explained that Breidenbach's T cells could be stimulated to recognize and attack melanoma. "Dr. Rosenberg basically told me to get on the plane on Monday and expect to be here for three weeks." Breidenbach's T cells had been removed and manipulated in Rosenberg's lab. Upon his arrival at the National Institutes of Health (NIH), they were returned to his body through a catheter entering the vein to his heart. "All the doctors were grinning in the operating room," he told me. "I felt like it was *Dr. Strangelove*." Breidenbach developed a fever of 104 degrees, and his skin erupted in a rash. He went home on Christmas Eve barely able to walk, but within a month the numerous metastases had started to shrink. Today none of the melanoma remains. "My T cells, they were fiery," Breidenbach concluded. But there was one permanent side effect of the treatment. Along with the cancer, the manipulated T cells attacked the normal cells with melanin, causing vitiligo, in which skin loses its pigment and hair turns white.

Rosenberg believes that melanoma has a unique relationship with the immune system: there are so many mutations in the tumors that T cells have an easier time recognizing them as foreign. This characteristic makes developing immune therapies easier. "An in-

tense natural immune response just doesn't exist for other kinds
of cancers," he said.

But Rosenberg thinks that he has the key to a more wide-rang-
ing approach. "With six hundred thousand Americans dying every
year with cancer, we need something for the common cancers," he
said. He acknowledges that targeted drugs, such as Gleevec, can
be effective, but he points out that most targeted therapies quickly
wane in their efficacy. A recently developed therapy for melanoma
dramatically shrank more than half of tumors, but nearly all patients relapsed within a year. A study published in March suggested
that as a cancer spreads in the body—from the kidney to the liver
and the lungs—the mutations occur in nonuniform ways, so that
DNA in liver deposits may differ from DNA in tumors in the lung.
This protean progression means that a drug targeted to one mutation may not work against cancer cells throughout the body.

In Rosenberg's view, with adoptive cell transfer, these malignancies would all appear equally foreign to the immune system. He
is refining the treatment for other cancers by skimming patients'
blood and then inserting a gene into their T cells that targets a different protein, called NY-ESO. The protein, which was identified
at Memorial Sloan-Kettering, is normally absent in tissues after
fetal development, except in the testis, but it reappears in about
a third of all common cancers. "I think adoptive cell transfer is
going to be the secret to applying immune therapy to the treatment of many human cancers," Rosenberg said. "When T cells are
genetically engineered to target NY-ESO, there is no difference
between melanoma and breast cancer or prostate cancer, or colon
cancer, ovarian cancer, sarcoma, and so on."

Varmus agrees that this approach might make a wider array of
tumors susceptible to therapy, and in early trials Rosenberg's strategy has been promising. In 2008 Anita Robertson, a sixty-three-
year-old accountant from Long Beach, California, had a large
sarcoma growing in her hip, a type of tumor similar to the one
that killed Elizabeth Dashiell. In July 2010, after treatment with genetically altered T cells, Robertson was discharged from the NIH
hospital. A CAT scan in September showed that the sarcoma had
begun to shrink; it is now more than 50 percent smaller. Once immobile and in pain from the cancer, she now can drive, shop, and
attend church.

Using a similar approach, researchers at the University of Pennsylvania have eradicated chronic lymphocytic leukemia in three patients who were no longer responding to other therapies. This month Rosenberg reported remissions in eight of nine patients with advanced lymphoma, and in three of those patients the cancer disappeared completely.

"We've got much, much better now with adoptive cell transfer," Rosenberg told me, "but it's not widely available." The treatment has to be individually designed for each patient, which makes it enormously expensive and so less valuable to pharmaceutical companies. "They want a drug, and they don't care if you spend five hundred million dollars developing the first vial, as long as they can produce the second vial for a dollar," Rosenberg said. Because his work is experimental, it has been supported by federal funds. Eventually, however, these therapies will be priced by calculating how much they offset the costs of conventional treatments. Although the new procedures could run to hundreds of thousands of dollars, they might still prove less costly than the money spent on chemotherapy, hospitalization, and hospice care for the many patients who currently cannot be cured.

Jedd Wolchok, however, argues that common cancers may not require adoptive cell therapy. He talks about the "three E's" in immune therapy: elimination, equilibrium, and escape. Therapy should aim for total elimination of the cancer, but "we need to think about immune-system equilibrium," in which the cancer, though present, does not grow or spread. After decades of frustration and failure in the clinic, most scientists are wary of predicting whether immune therapy will be able to completely cure the majority of cancer patients. Tumors have mutated to escape the effects of radiation, chemotherapy, and targeted agents; the body's immune responses may not be unique.

Though CTLA-4 is still the focus of much research, scientists have now identified at least five other inhibitors on T cells. Initial studies show that treatments directed at these inhibitors can shrink some of the most deadly tumors, including those of the lung and the colon. Mario Sznol, an oncologist at Yale, has conducted clinical trials with an antibody directed against one of the inhibitors, a protein called PD-1. "I believe that in the future we can customize immune therapy to the individual patient," he said.

Doctors will examine the specific characteristics of a tumor and then treat patients with the appropriate antibody.

Allison's laboratory is an open space that occupies a large part of the fifteenth floor of the Zuckerman Research Building at Sloan-Kettering. The day I visited, postdoctoral fellows and graduate students were analyzing data on their computers from recent experiments. In a corner was an intravital microscope, which can show cells and tissues in a living animal. Allison demonstrated how an anesthetized mouse is injected with the antibody to CTLA-4. Previously, the T cells of the mouse had been labeled with a fluorescein dye and sensitized to a protein from a tumor. Using the intravital microscope, "you can actually watch the T cells move into the lymph node," Allison said. They appeared as bright green circles coursing through thin gray vessels. "And then the T cells jump —they leave the lymph node and attack the tumor."

In another part of the lab, a postdoctoral fellow had arranged a series of mice that had been inoculated with melanoma. Some served as controls, and black masses an inch or more grew on their flanks. Others had received the antibody to CTLA-4 or to PD-1 or a combination. "The most dramatic regression is seen with the combination," Allison said, pointing to the flanks of mice where the tumors had shrunk to small black dots. Clinical trials in patients have begun with combining one antibody against CTLA-4 and another against PD-1 in order to remove two distinct brakes on the T cell.

Last year the antibody to CTLA-4, marketed under the name Yervoy, was approved by the Food and Drug Administration to treat melanoma. It was a vindication for immune therapy and an important step in the treatment of cancer. Yet this branch of research has also uncovered how far we have to go to understand the mutations that make cancer the most protean of diseases. "The future is about thoughtful combinations, different antibodies, perhaps with targeted therapies," Wolchok told me. "There won't be a single silver bullet for everyone."

DAVID OWEN

The Artificial Leaf

FROM *The New Yorker*

DANIEL NOCERA WAS a science-minded high-school junior in New Jersey at the beginning of the Arab oil embargo in 1973. American fuel prices soared, the stock market crashed, Congress prohibited speed limits higher than 55 miles an hour, and President Nixon banned the sale of gasoline on Sundays. At the end of the decade, the Iranian revolution, followed closely by the outbreak of war between Iran and Iraq, precipitated a second oil crisis. By then Nocera was a graduate student in chemistry at the California Institute of Technology. Within a short time, he had decided to devote his science career to energy.

Most of the energy we use comes from photosynthesis. Green plants store energy from the sun in chemical bonds, and we exploit that energy when we eat plants, or when we eat animals that have eaten plants, or when we burn either plants or substances ultimately derived from plants: firewood, peat, coal, oil, natural gas, ethanol. Photosynthesis has been understood in a general way for a long time and is familiar even to grade-school students—water and carbon dioxide in; oxygen and carbohydrates out—but the process is complex, and until fairly recently important parts of it remained mysterious. Nevertheless, Nocera decided in the early eighties that the chemistry of green plants was the likeliest place to seek an answer to civilization's long-term energy difficulties. "For the past two hundred years, we've run this other experiment, with fossil fuels, and it's not working out so well," he told me last August in his office in the chemistry department at the Massachusetts Institute of Technology. (Next January he will move to Harvard.)

"I wanted to go back to what worked for two billion years before that."

When the price of oil dropped in the mideighties, alternative-fuel research declined in popularity as an academic pursuit. "There weren't even conferences for me to go to," Nocera said, "because everybody had left the field." But he persisted in his research, seeking a way to inexpensively replicate solar-energy conversion as performed by vegetation. His early work focused on certain reactions that underlie key parts of that process and created a field now known as "proton-coupled electron transfer." In 2000 he decided that he understood the fundamental science well enough to, in his words, "go for it," and at the 2011 national meeting of the American Chemical Society, he announced a tangible breakthrough: a cheap, playing-card-size coated silicon sheet that, when placed in a glass of tap water and exposed to sunlight, split the water into hydrogen and oxygen. A video he made shows gas bubbles streaming from the sheet. He said the gas could easily be collected and either burned or used to power a fuel cell. He called the device an "artificial leaf."

Two months later, I heard Nocera give an hour-long presentation to a large audience of environmentalists, scientists, engineers, economists, government officials, and others at the Aspen Environment Forum. The talk was moderated by an editor of the British scientific journal *Nature*, who described Nocera's work as "the hydrogen economy reimagined" and said that Nocera's discoveries might constitute "an answer to the energy puzzle." Nocera claimed that artificial leaves could enable people everywhere to live without being connected to any power grid. He predicted that within a few decades, "you will be in control of your own energy," and that the artificial leaf would enable his audience to turn a home into "a self-sufficient power station." But he had a warning too: "Don't clap yet, because you don't want to do it."

Nocera's forebears are Italian, by way of Brooklyn and Medford, Massachusetts, but his Mediterranean profile and graying beard give him a rabbinical look. "I grew up in a very Jewish part of New Jersey," he told me. "To get back at my mother and father—who were Catholics, of course—I became the best Orthodox Jew in the world. I would go to temple on Saturday with my friends, and because I was doing it out of a passion for irritating my mother, I was a much better Jew than they were." At Michigan State, where he taught for thirteen years after earning a PhD at Caltech, his secretary was a Black Muslim, and he sometimes accompanied her to her mosque. Such ecumenism, he said, made him tolerant of all religions while killing any religious feelings of his own. He does, however, think of himself as a spiritual person, and he said that his spirituality has manifested itself mainly in a quest to provide low-cost, carbon-neutral fuels to the world, whose energy consumption he expects will double in the next forty years.

Nocera's father was a clothing buyer, first for Sears, then for J.C. Penney. Until Nocera reached eighth grade, his family moved so frequently that he made very few friends. "People ask me how I became a scientist, and this is literally how I did it," he told me. "When I was in grade school, I didn't invest in people. I would consciously not make friends, because I knew I would get attached to them and they would evaporate. The one thing I could bring with me when we moved was science." He loved Heathkits—moderately priced build-it-yourself gadgets that were popular with nerdy tinkerers—and when he was in second or third grade he used a microscope to conduct minutely detailed subsurface investigations of small portions of his backyard.

A second scientific milestone—although he didn't recognize it as such at the time—occurred when, as a teenager, he became an ardent fan of the Grateful Dead, and he sometimes left home for weeks to follow the group on tour. In Aspen last June, he had an animated discussion with two other conference participants about what one of them had identified as an overrepresentation of Deadheads in high-tech companies in Silicon Valley. The hypothesis was that a love for the Dead reflects an iconoclastic outlook that's conducive to innovative thinking, and that Deadheads share what Nocera now thinks of as a Garcian conception of open-source collaboration. "The Dead decentralized music," he said. "They would let you tape their shows, and back before there were computers and the Internet, there was this huge underground community that swapped recordings. I'd never really thought about it before Aspen, but what I want to do with energy is no different from what the Dead did with their music. I want to distribute it to everybody."

The process that Nocera calls "artificial photosynthesis" could be described more precisely as solar-powered electrolysis of water:

using energy from the sun to electrochemically split water into
hydrogen and oxygen. In natural photosynthesis, photons from
the sun, aided by chemicals in leaves, strip electrons from water,
breaking it into oxygen molecules and hydrogen nuclei. The plant
discards the oxygen—incidentally creating the part of the atmosphere that's the most important to us—and combines the hydrogen with atmospheric carbon dioxide to produce cellulose (the
material out of which the plant builds its structure) and starch
(the plant's fuel). "In terms of energy, those last parts are an afterthought," Nocera said. "The plant does them because it needs
to stand up and it can't deal with a gas. But all I care about is the
rearranged bonds." Natural photosynthesis is starkly inefficient:
many plants convert as little as 1 percent of the energy in the sunlight that falls on them. But sunlight is plentiful, and the energy in
the photons that strike the Earth each hour is roughly equivalent
to the total energy, from all sources, that humans use in a year. In
2009 Fred Krupp, who is the president of the Environmental Defense Fund, wrote that Nocera's work "makes it conceivable that
by midcentury we could satisfy our global energy needs by splitting
—each second—just a third of the water in MIT's swimming pool."

Electrolysis of water is sufficiently nonbaffling to be performed
routinely by schoolchildren: attaching electrodes to a battery and
placing them in water causes oxygen to bubble from one and hydrogen to bubble from the other. To Internet energy cranks, this
looks like a zero-carbon bonanza: free fuel! But hydrogen is really an energy carrier—a storage medium—rather than an energy
source, since burning or otherwise converting it merely releases
energy that was previously contained in the battery. Large quantities of hydrogen are used in industry, but the gas is too light
to exist on Earth naturally, and almost all the world's supply is
manufactured by doing things like steam-reforming natural gas,
gasifying coal, and partly oxidizing petroleum. The enthusiasm
that many environmentalists express for the (still hypothetical) hydrogen economy depends, first of all, on finding affordable ways
to produce hydrogen that don't add carbon to the atmosphere or
consume more fossil fuel than they replace. Hence Nocera's focus
on plants and the sun.

Nocera isn't the only scientist working on artificial photosynthesis. The field is at least four decades old, and interest in it has

grown in recent years. Some of the earliest research was done in Japan, which imports 90 percent of its oil and therefore has an incentive to find nonfossil alternatives, especially now that its zeal for nuclear reactors has weakened. In 2010 the U.S. Department of Energy and Caltech established the Joint Center for Artificial Photosynthesis, which has a five-year budget of $122 million. The center coordinates work by numerous researchers, who are exploring various approaches to solar-energy conversion. An important advance in artificial photosynthesis occurred in 1998 when John Turner—a scientist at the National Renewable Energy Laboratory, which is also funded by the Department of Energy—built a device that, like Nocera's, split water with no power source other than sunlight. Turner's device was extremely expensive and its components corroded rapidly, but before it stopped working it had converted approximately 12 percent of the energy in the sunlight that fell on it, making it up to twelve times better at fuel production than a green plant.

Nocera's artificial leaf, in its current form, is less than half as efficient as Turner's, but it's far more durable and only a fraction as expensive. Earlier this year, Nocera founded a company, Sun Catalytix, to pursue artificial photosynthesis, energy storage, and renewable fuels. Sun Catalytix has received almost all its funding from three sources: a technology-venture-capital firm, the Department of Energy's ARPA-E program, and the Indian industrialist and billionaire Ratan Tata. The company has designed an inexpensive water-splitting device, which the Tata Group expects to test on some scale in India late this year or early next year. "Mr. Tata and I spoke for only fifteen minutes the first time we met," Nocera told me. "But he kind of shares my vision, which is sun plus water is energy for the world."

On campus maps, MIT's buildings have assertively unimaginative designations. Nocera's office is in Building 6, which is situated between Building 8 and Building 2 and across a shady quadrangle from Building 18. His office is on the third floor. In the hallway outside his office door on a recent morning were seven large liquid-nitrogen canisters, like empties left for the milkman. "We're always making stuff," he explained, "and when you make stuff you need liquid nitrogen." (It's used to keep things very cold.) He led me into a laboratory on the other side of the hall and pointed out various pieces of equipment. "There's something in condensed-matter physics called quantum spin liquid," he said. "It's a very exotic state of matter, and we actually made it in this furnace. And that machine, over there, is a fifty-thousand-dollar microwave oven. One of the students in my group had the idea of cooking chemicals with microwaves, and that's what we do with it."

Nocera had just returned from an international conference on Lord Howe Island, a minimally populated six-mile-long volcanic outcropping in the Pacific Ocean several hundred miles northeast of Sydney. Because he travels so much, he spends less time than he would like working directly with the graduate students and post-doctoral fellows under his supervision. Nevertheless, he told me, he can usually recognize most of them. "That's Noémie," he said, to demonstrate. "She's from France. And that's Elizabeth, from Texas. And that's a random guy I don't know."

"I'm Oliver," the guy said.

"Oliver! From Germany!" (Oliver Bruns, who is from Hamburg, wasn't one of Nocera's graduate students but was visiting from an-other group in the chemistry department.)

We moved to a large glass-fronted cabinet. "This is a glove box," Nocera said. "A lot of the compounds we work with would catch on fire or explode in air, so this box has no oxygen in it. If the catalysts I use for artificial photosynthesis had to live in a world of no oxygen, I'd be in trouble, but the weird thing is that discover-ing how to make those catalysts often involved working with com-pounds that catch on fire and explode."

Every researcher who is interested in artificial water-splitting faces bigger challenges than a schoolroom demonstration of electrolysis might suggest. Most of those challenges arise because water is so chemically stable that its components resist molecular rearrangement—a good thing, since it prevents the spontaneous combustion of oceans, among other undesirable phenomena. For electrolysis to work, that molecular stubbornness has to be over-come—by turning up the voltage, by adding chemicals to the wa-ter, by using catalysts that promote the shuffling of bonds. But such measures also increase costs, and some of them reduce the longevity and reliability of the components. "In natural photosyn-thesis, the catalyst is the leaf," Nocera said, and much of his research has involved looking for affordable materials that reliably perform similar functions.

In 2008 Matthew W. Kanan, a postdoctoral fellow who was working with Nocera—he's now a chemistry professor at Stanford—was trying to find an improved catalyst for the oxygen-producing side of the water-splitting reaction. Nocera had successfully used compounds made of iridium and rhodium, but both metals are rare and expensive. "On the periodic table, elements that are next to each other usually have similar properties," Nocera said, "and if you look up from iridium and rhodium, in the same column, you see cobalt, which is stuff you find in rocks." Kanan prepared a cobalt compound and added it to the water in a test device, along with a phosphate buffer—to neutralize any acids formed in the reaction—and turned on the electric current. Before long, Kanan told me, a "golden-green layer" began to form on the surface of the electrode, and he assumed that his experiment had failed. He kept the power on, though, and after about an hour he saw a stream of bubbles. He and Nocera eventually determined that the bubbles were pure oxygen, and that the dissolved cobalt and phosphate had combined on their own to form a highly effective, low-cost catalyst, which now coated the electrode.

This was not just a chemical breakthrough but also a philosophical one. Researchers in the past, Nocera said, had focused on finding catalysts that didn't break down when they were submerged in water and exposed to sunlight, electric currents, chemical additives, and one another. (Platinum, iridium, and rhodium resist corrosion in almost any environment.) But Kanan's discovery suggested that an effective way to increase long-term reliability might be to use materials that re-formed after decomposing. This is a trick that nature employs throughout the biosphere. "A leaf doesn't have its own chemist—it has to assemble itself and heal itself," Nocera said. "We realized that if we had catalysts that repaired themselves we could achieve sort of the same thing, and that's what cobalt and phosphate have let us do."

Kanan's cobalt-phosphate catalyst, furthermore, deposits itself in a layer so thin that it's virtually transparent to sunlight. Nocera's earlier hydrogen-generating devices had been powered by external photovoltaic panels—the same solar panels you sometimes see on suburban roofs. These are made from materials—usually forms of silicon—that produce an electric current when they're exposed to light. The transparency of the cobalt-phosphate catalyst enabled Nocera, in effect, to move the solar panel into the water and to eliminate the wires. The artificial leaf, in its current version, is a piece of silicon coated on one side with Kanan's cobalt-phosphate compound, and on the other with an inexpensive nickel-based catalyst (which is required for the hydrogen side of the reaction). Nocera said, "You drop it in a glass of water and hold it up to the sun, and it starts generating fuel." When light penetrates the cobalt layer, a wireless electric current arises within the silicon, and the sandwich's two faces become electrodes, causing hydrogen to bubble from one and oxygen to bubble from the other. Kanan's catalyst, serendipitously, also allows the reaction to run in almost any water, including seawater and human wastewater—a huge advantage, since in much of the world pure water is scarcer than fossil fuel.

Whenever the artificial leaf is mentioned in a newspaper article or on the web, Nocera finds out almost immediately, because his e-mail and voice-mail inboxes fill up. Many of the messages come from young scientists who want to study with him. "The others are from people who say, 'Nocera, I heard you invented this technology. Please come install it at my house tomorrow,'" he said. The inquiries from students gratify him, because they suggest that the scientific world is coming around to his point of view. His reaction to the other inquiries is more complicated. For one thing, Nocera's artificial leaf hasn't evolved to the point where he or anyone else could install it at someone's house. For another thing, the people who contact him about buying the device are usually denizens of what Nocera calls "the legacy world"—the fortunate minority of the Earth's population that historically has enjoyed most of the considerable benefits of burning fossil fuels. These are not the people he views as the target users of his technology, at least in the near term. Since the early eighties, he has focused on the non-legacy world—the billions of impoverished people who have little or no access to modern fuels or to any electricity grid. "If there's one thing that's unique to the technology development I've done, it's been doing science with the super-poor in mind," he told me. His emphasis is largely humanitarian; it also arises from his belief, as a scientist, that the only way to meet the world's projected energy needs without causing intolerable environmental harm will be to work, in effect, from the bottom up — an approach that's very different from the ones that dominate energy research.

To contented inhabitants of the legacy world, mankind's gathering energy and climate challenges often seem to be primarily automotive. Most carbon-neutral energy research has focused on providing renewable energy in the way that cars and other power-hungry devices need it: in very large quantities, stored compactly and delivered quickly. Fossil fuels meet those requirements because they contain energy in a highly concentrated form. ("Every year, by burning fossil fuels, we release a million years of photosynthesis," Nocera told me.) Nonnuclear zero-carbon alternatives almost always lack this characteristic, which is known as "energy density." To use the sun to power an electric car, for example, you have to harvest light from an area that's much larger than the car's surface, and in order to travel long distances or at night you have to stockpile the energy in batteries, which are expensive and are incapable of storing more than about a hundredth as much energy as the same weight or volume of gasoline. The Tesla Roadster, an all-electric car, has a hand-assembled lithium-ion battery pack, which weighs 1,000 pounds, has a projected useful life of seven years, costs roughly $40,000 to replace, and takes a long time to recharge — many hours if the energy source is the low-voltage trickle from a photovoltaic panel. "There's always the promise that batteries are going to get better, but there's a physical ceiling on how closely you can pack electrons," Nocera said. "We could have chosen batteries instead of fuels a hundred years ago, but we didn't, and there's a reason. I like listening to people give talks about batteries as their laser pointer is running out of juice." The low energy density of batteries and sunlight helps to explain why the current proportion of total U.S. energy consumption that's supplied by photovoltaic panels, rounded to the nearest whole percentage point, is zero.

Hydrogen is extremely high in energy density; a pound contains almost three times as much energy as a pound of gasoline. But as a fuel, hydrogen has drawbacks, the most significant of which is that its physical density is extremely low. Storing an automotively useful quantity of hydrogen in a container the size of a car's gas tank requires enormous compression and expensive composite materials. Furthermore, most hydrogen-powered demonstration cars have engines called fuel cells, which are costly and technologically complex and which don't function at some temperatures. And, of course, the environmental benefit of using hydrogen as a fuel is lost if the hydrogen is manufactured from fossil fuels.

The low-carbon-energy challenge looks different, however, if you ignore SUVs and think of the billions of people who live in extreme poverty. For them, the main energy concern is not how to accelerate a 4,000-pound vehicle from a stop to highway speed in a few seconds and cruise for hundreds of miles; it's how to survive from one day to the next. For such people, Nocera told me, the main energy issue isn't power or efficiency or energy density; it's cost. "For the nonlegacy world, energy has to be super-cheap," he said. "If I could make alternative energy that was cheap enough for you to want to use it in your house—and I can't—it still wouldn't be cheap enough for the poor." Providing energy for these people has been Nocera's goal from the beginning. "In the next forty years, three hundred and fifty million Indians are going to become energy users," he said. "We've got to get them energy, and it's got to be CO_2-neutral, because if they use coal we're screwed."

Nocera's vision for the world's poorest people is of a gridless, decentralized energy system, in which every dwelling has an artificial leaf on its roof. When the sun shines, the leaf splits water—about a liter and a half per day—and after dark the residents burn the hydrogen in an inexpensive microturbine, which generates electricity till dawn at an average rate of about a hundred watts. By legacy-world standards, this is a truly minimal power level, but it's sufficient, Nocera thinks, to transform the lives of people who currently have none, or almost none. And it's cheap. The components are not particularly efficient, but they are low-tech and commercially available today. And because the fuel is produced in small quantities and used onsite, the hydrogen can be stored in ordinary metal tanks, at modest pressure. Furthermore, the self-repairing cobalt-phosphate catalyst keeps the need for maintenance low—a critical factor, Nocera said, because "you can't have a bunch of people running around the world fixing stuff."

Nocera believes that the benefits of large-scale implementation would extend beyond direct, energy-related gains in users' quality of life (illumination, cooking without burning wood, telephone charging) and would include a global decline in the rate of population growth, which, historically, has slowed as affluence has risen. "The real issue driving our problems on the face of this planet is population," he said. "One of the beautiful things about providing distributed energy to the poor is that it's a positive feedback loop. If I give poor people energy, they become empowered, and every study that's ever been done has shown that with financial gain and education population drops like a rock."

It's usually argued that complex technological gains trickle down to the poor—that the innovations required to reduce the sticker price of a Tesla Roadster from $110,000 to $80,000 will also eventually improve the lives of people at the bottom of the global income scale. But there's reason to think otherwise: as energy technology has grown in both sophistication and efficiency, the worldwide gap between richest and poorest has widened, and the richest countries today often treat the poorest ones less as partners in progress than as cheap targets for resource extraction. Nocera believes that simple technology scales up more readily than complex technology scales down. "The poor are helping you," he told his audience in Aspen, "because they're going to teach you how to live for the future."

Matthew Kanan, the Stanford chemistry professor who discovered Nocera's cobalt-phosphate catalyst, told me, "Dan isn't the only one who has made this point, but he's right that the developing world's energy trajectory is the one that's the most important over the next several decades." Focusing on people whose energy consumption is tightly constrained also reduces the likelihood of certain kinds of unintended consequences. A seldom discussed environmental danger posed by electric cars, for example, is that broad, rapid adoption would hugely increase, rather than reduce, demand for grid-supplied electricity generated by burning fossil fuels, since growth in renewable sources couldn't conceivably keep up. ("I totally hate the electric car," Nocera told me.)

Providing decentralized energy to the developing world carries a threat of unintended environmental consequences, too, of course. One possibility is that the artificial leaf could turn out to be the energy equivalent of a gateway drug. Historically, more energy has always meant more income, and more income has meant more consumption, and more consumption has meant more energy in every form—as well as increased demand for a rapidly expanding list of environmentally destructive possessions, including, eventually, the ultimate modern consumer good, the automobile. And although distributed energy production eliminates the need for a centralized electricity grid, it encourages the creation and enlargement of other environmentally problematic grids, including the ones used by phone calls, websites, food producers, airplanes, delivery trucks, and cars.

Among some scientists, Nocera has a reputation for hyping his discoveries. The playing-card-size device truly does split water, but it's a prop, not a product, since producing enough hydrogen to meet even Nocera's minimal goal of powering a single 100-watt light bulb through the night would require an artificial leaf the size of a door. Photovoltaic panels have the same size constraint, which arises from the diffuseness of sunlight and from silicon's ultimately limited ability to absorb it. "One thing the layperson messes up is that you can't go faster than the sun gives out energy," Nocera said. The advantage of the artificial leaf is not that it converts more solar energy than a conventional photovoltaic panel. The advantage is that it stores solar energy in a fuel rather than in a battery and is therefore potentially more versatile, as well as being less expensive to acquire, maintain, and exploit—as long as users' energy requirements are minimal.

Nocera's claims have also often been amplified by reporters, and even by his own university's public relations office. He hasn't always rushed to correct misimpressions, and at least some of his overselling has been intentional. Attracting funding for renewable-energy research requires showmanship, and the need for shrewd marketing has grown in recent years, as legacy-world interest in carbon-free energy has slackened. A further difficulty is that the science of renewable energy is genuinely daunting. Nocera's challenge outside the laboratory has been to build enthusiasm for the artificial leaf even though, in anything like its current form, it is designed to meet a level of energy demand that by modern American standards is almost immeasurably low.

As Nocera concedes, artificial photosynthesis, if it turns out to be practical for anyone, is almost certainly decades from large-scale implementation—a discouraging fact that applies to virtually all renewables. Still, he believes that if scientists and engineers were to apply the kind of effort to developing low-cost fuel cells for third-world homes that they now apply to developing high-performance batteries for American sports cars, they might accomplish something globally significant. They might also eventually find an economical way to replicate the far more challenging second stage of photosynthesis, in which hydrogen and atmospheric carbon dioxide combine to make a nongaseous fuel—a breakthrough that would eliminate the problems associated with storing and transporting hydrogen. "If we all just focused on this, in a coordinated way, I'm sure the science and engineering community could nail it," he told me. For that reason, he doesn't mind speculating publicly about outcomes that, realistically, he can't deliver yet. "A lot of scientists get mad at me for speaking at things like this," he said in Aspen, "because they think I'm going to give you hope."

MICHAEL SPECTER

The Deadliest Virus

FROM *The New Yorker*

ON MAY 21, 1997, a three-year-old boy died in Hong Kong from
a viral infection that turned out to be influenza. The death was
not unusual: flu viruses kill hundreds of thousands of people every
year. Hong Kong is among the world's most densely populated cities, and pandemics have a long history of first appearing there or
in nearby regions of southern China and then spreading rapidly
around the globe.

This strain, however, was unusual, and it took an international
team of virologists three months to identify it as H5N1 — "bird flu,"
as it has come to be called. Avian influenza had been responsible
for the deaths of hundreds of millions of chickens, but there had
never been a report of an infected person, even among poultry
workers.

By the end of the year, eighteen people in Hong Kong had become sick, and six had died. That's a remarkably high mortality
rate: if seasonal flu were as virulent, it would kill 20 million Americans a year. Hong Kong health officials, fearing that the virus was
on the verge of becoming extremely contagious, acted forcefully
to build a moat around the outbreak: during the last week of December, they destroyed every chicken in the city.

The tactic worked. Bird flu disappeared, at least for a while.
"We felt we had dodged a bullet," Keiji Fukuda told me earlier this
year when I visited him in his office at the World Health Organization's headquarters in Geneva. Fukuda, as the assistant director-
general for health, security, and environment, oversees influenza
planning. At the end of 1997, when he was the chief influenza

epidemiologist at the Centers for Disease Control and Prevention in Atlanta, he spent a few tense weeks in Hong Kong, searching for clues to how the virus was transmitted from chickens to humans and whether it would set off a global pandemic. "It was a very scary time," he said, "and we were bracing ourselves for the worst. But by the end of the month nobody else got sick, so we crossed our fingers and went back to Atlanta."

Then in 2003, the virus reemerged in Thailand; it has since killed 346 of the 587 people it is known to have infected—nearly 60 percent. The true percentage is undoubtedly lower, since many cases go unreported. Even so, the Spanish flu epidemic of 1918, which killed at least 50 million people, had a mortality rate of between 2 and 3 percent. Influenza normally kills far fewer than one-tenth of 1 percent of those infected. This makes H5N1 one of the deadliest microbes known to medical science.

To ignite a pandemic, even the most lethal virus would need to meet three conditions: it would have to be one that humans hadn't confronted before, so that they lacked antibodies; it would have to kill them; and it would have to spread easily—through a cough, for instance, or a handshake. Bird flu meets the first two criteria but not the third. Virologists regard cyclical pandemics as inevitable; as with earthquakes, though, it is impossible to predict when they will occur. Flu viruses mutate rapidly, but over time they tend to weaken, and researchers hoped that this would be the case with H5N1. Nonetheless, for the past decade the threat of an airborne bird flu lingered ominously in the dark imaginings of scientists around the world. Then, last September, the threat became real.

At the annual meeting of the European Scientific Working Group on Influenza, in Malta, several hundred astonished scientists sat in silence as Ron Fouchier, a Dutch virologist at the Erasmus Medical Center in Rotterdam, reported that simply transferring avian influenza from one ferret to another had made it highly contagious. Fouchier explained that he and his colleagues "mutated the hell out of H5N1"—meaning that they had altered the genetic sequence of the virus in a variety of ways. That had no effect. Then, as Fouchier later put it, "someone finally convinced me to do something really, really stupid." He spread the virus the old-fashioned way, by squirting the mutated H5N1 into the nose of a ferret and then implanting nasal fluid from that ferret into the nose of another. After ten such manipulations, the virus began to spread around the ferret cages in his lab. Ferrets that received high doses of H5N1 died within days, but several survived exposure to lower doses.

When Fouchier examined the flu cells closely, however, he became alarmed. There were only five genetic changes in two of the viruses' eight genes. But each mutation had already been found circulating naturally in influenza viruses. Fouchier's achievement was to place all five mutations together in one virus, which meant that nature could do precisely what he had done in the lab. Another team of researchers, led by Yoshihiro Kawaoka, at the University of Wisconsin, created a slightly different form of the virus, which, while not as virulent, was also highly contagious. One of the world's most persistent horror fantasies, expressed everywhere from Mary Shelley's *Frankenstein* to *Jurassic Park*, had suddenly come to pass: a dangerous form of life, manipulated and enhanced by man, had become lethal.

Fouchier's report caused a sensation. Scientists harbored new fears of a natural pandemic, and biological-weapons experts maintained that Fouchier's bird flu posed a threat to hundreds of millions of people. The most important question about the continued use of the virus, and the hardest to answer, is how likely it is to escape the laboratory. "I am not nearly as worried about terrorists as I am about an incredibly smart, smug kid at Harvard, or a lone crazy employee with access to these sequences," Michael T. Osterholm, the director of the Center for Infectious Disease Research and Policy at the University of Minnesota Health Center, told me. Osterholm is one of the nation's leading experts on influenza and bioterrorism. "We have seen many times that accidental releases of dangerous microbes are not rare," he said.

Osterholm's anxiety was based in recent history. The last person known to have died of smallpox, in 1978, was a medical photographer in England named Janet Parker, who worked in the anatomy department of the University of Birmingham Medical School. Parker became fatally ill after she was accidentally exposed to smallpox grown in a research lab on the floor below her office. In the late 1970s, a strain of H1N1—"swine flu"—was isolated in northern China near the Russian border, and it later spread throughout the world. Most virologists familiar with the outbreak are convinced that it came from a sample that was frozen in a lab and then released accidentally. In 2003 several laboratory technicians in Hong Kong were infected with the SARS virus. The following year, a Russian scientist died after mistakenly infecting herself with the Ebola virus.

Biological labs are given four possible biosafety-level security grades, ranging from BSL-1 to BSL-4. Research on the most lethal and contagious organisms is carried out at BSL-4 laboratories. Under U.S. guidelines, BSL-3 facilities contain microbes that cause "serious or potentially lethal diseases" but do not easily pass among people or for which there are easily accessible preventives. BSL-4 laboratories house agents that have no preventives or treatments. The labs in Rotterdam and in Wisconsin where the H5N1 ferret work was conducted were both BSL-3 facilities that had been enhanced with additional security measures. In such laboratories, scientists are typically subjected to security checks; they wear space suits and breathe through special respirators. Although no safeguards are absolute, negative air filters attempt to ensure that no particles accidentally escape from the lab.

Last December the National Science Advisory Board for Biosecurity, a panel of science, defense, and public-health experts, was asked by the Department of Health and Human Services to evaluate Fouchier's research. The panel recommended that the two principal scientific journals, *Science* and *Nature*, reconsider plans to publish information about the methods used to create the H5N1 virus. It was the first time that the advisory board, which was formed after the anthrax attacks of 2001 to provide guidance on "dual use" scientific research, which could both harm and protect the public, had issued such a request. "We are in the midst of a revolutionary period in the life sciences," the advisers wrote. "With this has come unprecedented potential for better control of infectious diseases and significant societal benefit. However, there is also a growing risk that the same science will be deliberately misused and that the consequences could be catastrophic." The *New York Times* published an editorial that echoed the advisory board's concern and even questioned the purpose of the experiments: "We believe in robust research and almost always oppose censorship. But in this case the risks—of doing the work and publishing the results—far outweigh the benefits." The journal *New Scientist* agreed: ONE MISTAKE AWAY FROM A WORLDWIDE FLU

PANDEMIC. Television talk shows and the Internet pulsated with anxiety.

The widespread alarm led *Science* and *Nature* to agree to postpone publication. Fouchier's virus, which now sits in a vault within his securely guarded underground laboratory in Rotterdam, has fundamentally altered the scope of the biological sciences. Like the research that led to splitting the atom and the creation of nuclear energy, the knowledge that his experiment has provided could be used to attack the public as well as to protect it.

"Terror is not an unjustified reaction to knowing this virus exists," Osterholm, who serves on the advisory board, told me. "We have no room to be wrong about this. None. We can be wrong about other things. If smallpox got out, it would be unfortunate, but it has a fourteen-day incubation period, it's easy to recognize, and we would stop it. Much the same is true with SARS. But with flu you are infectious before you even know you are sick. And when it gets out it is gone. Those researchers have all of our lives at the ends of their fingers."

Fouchier, a lanky forty-five-year-old man with intense blue eyes, works at one of the most highly regarded virological laboratories in Europe. "I have spent many years, and this institution has paid millions of dollars to insure that this research was carried out in the safest possible manner," he told me when we met in a conference room in the grim research facility that houses his laboratories at the Erasmus Medical Center. The center devoted several years to constructing a special lab for Fouchier's research. From the windows, one can see barges and hulking gray cranes; Rotterdam is Europe's busiest port. It is an industrial cityscape whose bleakness, on the day I visited, seemed to match Fouchier's mood. As he spoke, he stared at his hands, which he clenched nervously. "People are acting like I am some mad scientist," he said.

Fouchier spent much of his career working on the structure of the AIDS virus. In 1997 he abruptly turned to bird flu, both because he was fascinated by its molecular structure and because he quickly grasped its pandemic potential. He has published scores of scientific articles on how influenza viruses move between species. Since December, however, when the advisory board recommended postponing publication of the bird-flu research, and some of his colleagues called for stopping it entirely, he has felt, he says, like the focus of "an international witch hunt." He was incensed. "To attempt to prevent this research from reaching the largest number of scientists is bullshit," he told me. "The more people who have access to it, the more likely we are to get answers to the many questions we still need to ask. Everyone who knows anything about virology can get hold of the recipe." There were nearly a thousand people at the Malta meeting where he first announced his findings. "This moratorium serves some fake sense of security," he said. "It does not serve the public health."

Fouchier, as well as Kawaoka and other researchers, had been trying for years to learn whether H5N1 could trigger a worldwide pandemic. He wondered why the virus has destroyed so many poultry flocks in the United States, Europe, and Asia but infected so few people. Fouchier hoped to characterize the properties that make the virus so much deadlier than others. The only way to answer these questions was to create a variant that would cling to human cells in the nose and throat. Fouchier's research was hardly the work of a furtive renegade. Several international review committees oversaw his experiments, and he received funding from the National Institutes of Health. Despite the risks, most people in his field believed that the experiments were necessary. Moreover, they were not without precedent. In 2002 Eckard Wimmer, at Stony Brook University, stitched together hundreds of DNA fragments, mostly acquired via the Internet, then used them to create a fully functional polio virus. In the fall of 2005, several published academic papers described the genomic sequence of the 1918 Spanish flu, which caused the world's deadliest influenza pandemic. In each case, the publications were initially denounced but were eventually accepted as valuable.

"In this profession, you always do it wrong," Ab Osterhaus, a leading infectious-disease expert who runs the virology department at Erasmus, said. "Either you give too much warning or not enough. Either you take things too seriously or not seriously enough. Fouchier's work is essential, and the questions it raises must be addressed."

There have been many hypotheses about how bird flu could become epidemic. Most researchers had believed that the avian virus would have to combine with human genes in pigs. Pigs usually serve as a mixing vessel for influenza viruses that make the transition from poultry to humans. (This is how the global pandemic starts in Steven Soderbergh's recent film *Contagion:* Gwyneth Paltrow is exposed to a pig that's been infected by a bat, and soon much of the world is dead or dying.) Other scientists believed that the H5 protein, because of its molecular structure, could not easily infect human cells. (Strains of influenza are named for two proteins on their surface that latch on to respiratory cells and make it possible for them to invade our lungs.) "There has been a lot of speculation that this virus cannot be transmitted easily or through the air," Fouchier told me. "That speculation has been wrong."

Although no animal study can predict with certainty what will happen in humans, ferrets get flu pretty much the way we do. Their lung physiology is similar to humans', and avian-influenza viruses bind to the same receptor cells in their respiratory tracts. Still, there has been sharp debate among scientists about whether results in ferrets can predict how humans will react to similar infections, with some researchers discounting the data entirely.

"The mutations . . . could cause the viruses to be more transmissible between humans," Peter Palese, a prominent microbiologist at Mount Sinai School of Medicine, wrote recently. "But this is simply unknowable from available data." Palese argues that the virus may be better adapted to ferrets than to other mammals.

"You cannot say, 'Just forget about it, because it happened in a ferret,'" Fouchier said. "This is our best model. But you also can't say, 'Because it happened in a ferret, it will happen in a human.' So it becomes a question of whether it's worth the risk of finding out. This is one of the most dangerous viruses you can imagine. It's not my virus—it's our virus. And it's out there. We need to deal with that. And if we focus on what matters, we can."

Once you create a virus that could kill millions of people, what should you do with it? And how should you handle the knowledge that made it possible?

There have been angry calls for Fouchier's virus to be destroyed, for it to be transferred to a military-level bioweapons facility, and for research to be stopped entirely. "It's just a bad idea for scientists to turn a lethal virus into a lethal and highly contagious virus," Dr. Thomas Inglesby, a bioterrorism expert and the director of the Center for Biosecurity at the University of Pittsburgh Medical Center, said. "And it's a second bad idea for them to publish how they did it, so others can copy it."

Still, most scientists who work with viruses insist that the value of this research outweighs the risks. Anthony S. Fauci, the long-time chief of the Institute of Allergy and Infectious Diseases, told me, "Those data could help scientists determine rapidly whether existing vaccines or drugs are effective against such a virus, as well as help in the development of new medications. It's hard to stop something if you don't know what it's made of. Naturally, if epidemiologists in countries where pandemics most often arise know what they are looking for, they will be able to move with greater urgency to contain the spread."

How likely is it that publishing the genetic sequence could help a terrorist, a rogue, or a legitimate researcher who might develop a novel vaccine or drug? "Most of us are unequivocal about the value of the research," Fauci said. "But deciding what to do with these types of studies is complicated. At the moment, there are no official governing bodies to regulate such decisions. They rely on the goodwill of researchers." Fauci and others have noted that precisely because flu is so hard to control, the virus would be difficult to use as a weapon.

In this case, as in most other cases, the work was supported heavily by the National Institutes of Health, and it seems unlikely to proceed without U.S. government support. Scientists bicker as vigorously as any other group, but rarely about the right to share and publish the data on which their research depends. Even the National Science Advisory Board for Biosecurity has made clear its general support for open investigation and full publication. The scientific method and the entire edifice of institutional research depend on such openness; without it, progress would slow dramatically. As biology has become more accessible, the balance between freedom and protection has become harder to maintain. This is certainly not the last time that preventing wide dissemination of information may seem necessary. But who should make those decisions, and how? Scientists fear that any regulatory body will stifle research. In 1975, when biologists met at Asilomar, California, to discuss the potential hazards of the new field of recombinant DNA technology, the group drew up voluntary guidelines to govern their research. Those guidelines have worked well, and that meeting is often regarded as a model of cooperative regulation.

We live in a very different world now. Secretary of State Hillary

Clinton recently gave a speech at a biological-weapons conference in Geneva in which she stressed that the threat of biological terror can no longer be ignored. "There are warning signs," Clinton said, including "evidence in Afghanistan that . . . Al Qaeda in the Arabian Peninsula made a call to arms for—and I quote—'brothers with degrees in microbiology or chemistry to develop a weapon of mass destruction.'"

While scientists disagree sharply about whether it would be easy to replicate such a virus in a laboratory, and whether it would be worth the effort, there is no question that we are moving toward a time when work like this, and even more complex biology, will be accessible to anyone with the will to use it, a few basic chemicals, and a relatively small amount of money.

Those realities have compelled many scientists to reconsider their unilateral support of the principle of open research. "I can tell you that when I began this journey I was certainly of the view that everything should be out and science should not be interfered with," Arturo Casadevall, the chief of infectious disease at the Albert Einstein College of Medicine and a member of the advisory board, said at a recent forum on the issue sponsored by the New York Academy of Sciences. "And as the result of hundreds of hours of the deliberative process I changed my mind." Others are even more emphatic, arguing that although the information is bound to become available, any delay is better than none. Many countries lack proper surveillance capacities, and existing vaccines are not good enough to stop influenza viruses from taking hold in the human population. By the time that public-health officials were fully aware of the swine-flu virus that originated in Mexico in 2009, for instance, it had spread across the globe.

In January, a few days before we met in Rotterdam, Fouchier had agreed to a sixty-day moratorium on the project, but only after he received a long, late-night phone call from Fauci, who convinced him that a worldwide timeout—the first since the beginning of the era of molecular biology—would allow people to cool off and enable them to explain the value of such research to the public. In mid-February, a committee of specialists, including Fouchier, met in Geneva at the WHO headquarters and announced that the papers would eventually be published in full, but that a sixty-day moratorium was probably not long enough. It is not clear when or where the research will continue.

Attempts to control information or to prohibit research rarely succeed for long. As the physicist and synthetic biologist Rob Carlson has written, most notably in his 2010 book, *Biology Is Technology*, in the case of crystal methamphetamine both prohibition and efforts by the federal government to shut down production labs have failed, and in similar ways. In each case, success in cracking down on small-time dealers led to failure on a larger scale. Carlson believes that cutting the flow of H5N1 data will have the same effect. "Any attempt to secure the data would have to start with an assessment of how widely it is already distributed," he wrote recently on his blog, Synthesis. "I have yet to meet an academic who regularly encrypts e-mail, and my suspicion is that few avail themselves of the built-in encryption on their laptops." Carlson noted that in addition to university computers and e-mail servers in facilities where the science originated, the information is probably stored in the computers of reviewers, on servers at *Nature* and *Science*, at the advisory board, and, depending on how the papers were distributed and discussed by the board's members, possibly on their various e-mail servers and individual computers as well. "And," Carlson wrote, "let's not forget the various unencrypted phones and tablets all of those reviewers now carry around."

Carlson and others argue that restricting publication would retard the progress of the research without increasing safety. With influenza viruses, speed matters. Vaccine production methods have not changed substantially in sixty years, and it was months before a useful vaccine was widely available for the H1N1 pandemic of 2009. That virus infected more than a billion people. Future bird-flu research could help scientists learn how it is transmitted through the air, why it makes the leap from animal to man, and how specifically it binds to human cell receptors. By placing the virus into tissue culture, scientists could discover more about how it destroys cells and make a better assessment of whether current vaccines would protect us—and if they wouldn't, the research could guide us toward making more effective vaccines. None of these experiments are without risk, but one must also consider the risk of not carrying them out.

"We can learn a great deal about transmission of influenza virus through the air from this work, and it's something we know very little about," Ab Osterhaus, the leader of the Erasmus team, said. "Nobody was going to make this virus in his garage. There are so many better ways to create terror. You have to compare the risk posed by nature with the theoretical risk that a human might use this virus for harm. I take the bioterror threat very seriously. But we have to address the problems logically. And nature is much more sophisticated than anyone in any lab. Nature is going to manufacture this virus or something like it. We know that. Bioterrorists might, but nature will. Look at the past century: the 1918 flu, HIV, Ebola, and H1N1. The Spanish flu took months. SARS, maybe a couple of weeks. This is happening all the time, and we have ways to fight it. So where is the greatest risk? Is it in someone's garage or in nature? Because you cannot prevent scientists from getting the information they need to address that risk. I understand politics and publicity. But I also understand that viruses do not care about any of that."

ALAN LIGHTMAN

Our Place in the Universe

FROM *Harper's Magazine*

MY MOST VIVID encounter with the vastness of nature occurred years ago on the Aegean Sea. My wife and I had chartered a sailboat for a two-week holiday in the Greek islands. After setting out from Piraeus, we headed south and hugged the coast, which we held three or four miles to our port. In the thick summer air, the distant shore appeared as a hazy beige ribbon—not entirely solid, but a reassuring line of reference. With binoculars, we could just make out the glinting of houses, fragments of buildings.

Then we passed the tip of Cape Sounion and turned west toward Hydra. Within a couple of hours, both the land and all other boats had disappeared. Looking around in a full circle, we could see only water, extending out and out in all directions until it joined with the sky. I felt insignificant, misplaced, a tiny odd trinket in a cavern of ocean and air.

Naturalists, biologists, philosophers, painters, and poets have labored to express the qualities of this strange world that we find ourselves in. Some things are prickly, others are smooth. Some are round, some jagged. Luminescent or dim. Mauve-colored. Pitter-patter in rhythm. Of all these aspects of things, none seems more immediate or vital than *size*. Large versus small. Consciously and unconsciously, we measure our physical size against the dimensions of other people, against animals, trees, oceans, mountains. As brainy as we think ourselves to be, our bodily size, our bigness, our simple volume and bulk are what we first present to the world. Somewhere in our fathoming of the cosmos, we must keep a men-

tal inventory of plain size and scale, going from atoms to microbes
to humans to oceans to planets to stars. And some of the most impressive additions to that inventory have occurred at the high end.
Simply put, the cosmos has gotten larger and larger. At each new
level of distance and scale, we have had to contend with a different
conception of the world that we live in.

The prize for exploring the greatest distance in space goes to a
man named Garth Illingworth, who works in a ten-by-fifteen-foot
office at the University of California, Santa Cruz. Illingworth studies galaxies so distant that their light has traveled through space
for more than 13 billion years to get here. His office is packed
with tables and chairs, bookshelves, computers, scattered papers,
issues of *Nature,* and a small refrigerator and a microwave to fuel
research that can extend into the wee hours of the morning.

Like most professional astronomers these days, Illingworth
does not look directly through a telescope. He gets his images
by remote control—in his case, quite remote. He uses the Hubble Space Telescope, which orbits Earth once every ninety-seven
minutes, high above the distorting effects of Earth's atmosphere.
Hubble takes digital photographs of galaxies and sends the images to other orbiting satellites, which relay them to a network of
earthbound antennae; these, in turn, pass the signals on to the
Goddard Space Flight Center in Greenbelt, Maryland. From there
the data is uploaded to a secure website that Illingworth can access
from a computer in his office.

The most distant galaxy Illingworth has seen so far goes by
the name UDFj-39546284 and was documented in early 2011.
This galaxy is about 100,000,000,000,000,000,000,000 miles away
from Earth, give or take. It appears as a faint red blob against the
speckled night of the distant universe—red because the light has
been stretched to longer and longer wavelengths as the galaxy has
made its lonely journey through space for billions of years. The
actual color of the galaxy is blue, the color of young, hot stars, and
it is twenty times smaller than our galaxy, the Milky Way. UDFj-
39546284 was one of the first galaxies to form in the universe.

"That little red dot is hellishly far away," Illingworth told me
recently. At sixty-five, he is a friendly bear of a man, with a ruddy
complexion, thick strawberry-blond hair, wire-rimmed glasses, and

a broad smile. "I sometimes think to myself: What would it be like to be out there, looking around?"

One measure of the progress of human civilization is the increasing scale of our maps. A clay tablet dating from about the twenty-fifth century B.C., found near what is now the Iraqi city of Kirkuk, depicts a river valley with a plot of land labeled as being 354 iku (about 30 acres) in size. In the earliest recorded cosmologies, such as the Babylonian *Enuma Elish*, from around 1500 B.C., the oceans, the continents, and the heavens were considered finite, but there were no scientific estimates of their dimensions. The early Greeks, including Homer, viewed Earth as a circular plane with the ocean enveloping it and Greece at the center, but there was no understanding of scale. In the early sixth century B.C., the Greek philosopher Anaximander, whom historians consider the first mapmaker, and his student Anaximenes proposed that the stars were attached to a giant crystalline sphere. But again there was no estimate of its size.

The first large object ever accurately measured was Earth, accomplished in the third century B.C. by Eratosthenes, a geographer who ran the Library of Alexandria. From travelers, Eratosthenes had heard the intriguing report that at noon on the summer solstice, in the town of Syene, due south of Alexandria, the sun casts no shadow at the bottom of a deep well. Evidently the sun is directly overhead at that time and place. (Before the invention of the clock, noon could be defined at each place as the moment when the sun was highest in the sky, whether that was exactly vertical or not.) Eratosthenes knew that the sun was not overhead at noon in Alexandria. In fact, it was tipped 7.2 degrees from the vertical, or about one-fiftieth of a circle—a fact he could determine by measuring the length of the shadow cast by a stick planted in the ground. That the sun could be directly overhead in one place and not another was due to the curvature of Earth. Eratosthenes reasoned that if he knew the distance from Alexandria to Syene, the full circumference of the planet must be about fifty times that distance. Traders passing through Alexandria told him that camels could make the trip to Syene in about 50 days, and it was known that a camel could cover 100 stadia (almost 11½ miles) in a day. So the ancient geographer estimated that Syene and Alexandria were about 570 miles apart. Consequently, the complete circumference of Earth he figured to be about 50 × 570 miles, or 28,500 miles. This number was within 15 percent of the modern measurement, amazingly accurate considering the imprecision of using camels as odometers.

As ingenious as they were, the ancient Greeks were not able to calculate the size of our solar system. That discovery had to wait for the invention of the telescope, nearly 2,000 years later. In 1672 the French astronomer Jean Richer determined the distance from Earth to Mars by measuring how much the position of the latter shifted against the background of stars from two different observation points on Earth. The two points were Paris (of course) and Cayenne, French Guiana. Using the distance to Mars, astronomers were also able to compute the distance from Earth to the sun, approximately 100 million miles.

A few years later, Isaac Newton managed to estimate the distance to the nearest stars. (Only someone as accomplished as Newton could have been the first to perform such a calculation and have it go almost unnoticed among his other achievements.) If one assumes that the stars are objects similar to our sun, equal in intrinsic luminosity, Newton asked, how far away would our sun have to be in order to appear as faint as nearby stars? Writing his computations in a spidery script, with a quill dipped in the ink of oak galls, Newton correctly concluded that the nearest stars are about 100,000 times the distance from Earth to the sun, about 10 trillion miles away. Newton's calculation is contained in a short section of his *Principia* titled simply "On the distance of the stars."

Newton's estimate of the distance to nearby stars was larger than any distance imagined before in human history. Even today, nothing in our experience allows us to relate to it. The fastest most of us have traveled is about 500 miles per hour, the cruising speed of a jet. If we set out for the nearest star beyond our solar system at that speed, it would take us about 5 million years to reach our destination. If we traveled in the fastest rocket ship ever manufactured on Earth, the trip would last 100,000 years, at least a thousand human life spans.

But even the distance to the nearest star is dwarfed by the measurements made in the early twentieth century by Henrietta Leavitt, an astronomer at the Harvard College Observatory. In

1912 she devised a new method for determining the distances to faraway stars. Certain stars, called Cepheid variables, were known to oscillate in brightness. Leavitt discovered that the cycle times of such stars are closely related to their intrinsic luminosities. More luminous stars have longer cycles. Measure the cycle time of such a star and you know its intrinsic luminosity. Then, by comparing its intrinsic luminosity to how bright it appears in the sky, you can infer its distance, just as you could gauge the distance to an approaching car at night if you knew the wattage of its headlights. Cepheid variables are scattered throughout the cosmos. They serve as cosmic distance signs in the highway of space.

Using Leavitt's method, astronomers were able to determine the size of the Milky Way, a giant congregation of about 200 billion stars. To express such mind-boggling sizes and distances, twentieth-century astronomers adopted a new unit called the light-year, the distance that light travels in a year—about 6 trillion miles. The nearest stars are several light-years away. The diameter of the Milky Way has been measured at about 100,000 light-years. In other words, it takes a ray of light 100,000 years to travel from one side of the Milky Way to the other.

There are galaxies beyond our own. They have names like Andromeda (one of the nearest), Sculptor, Messier 87, Malin 1, IC 1101. The average distance between galaxies, again determined by Leavitt's method, is about twenty galactic diameters, or 2 million light-years. To a giant cosmic being leisurely strolling through the universe and not limited by distance or time, galaxies would appear as illuminated mansions scattered about the dark countryside of space. As far as we know, galaxies are the largest objects in the cosmos. If we sorted the long inventory of material objects in nature by size, we would start with subatomic particles like electrons and end up with galaxies.

Over the past century, astronomers have been able to probe deeper and deeper into space, looking out to distances of hundreds of millions of light-years and farther. A question naturally arises: Could the physical universe be unending in size? That is, as we build bigger and bigger telescopes sensitive to fainter and fainter light, will we continue to see objects farther and farther away—like the third emperor of the Ming Dynasty, Yongle, who surveyed his new palace in the Forbidden City and walked from room to room to room, never reaching the end?

Here we must take into account a curious relationship between distance and time. Because light travels at a fast (186,000 miles per second) but not infinite speed, when we look at a distant object in space we must remember that a significant amount of time has passed between the emission of the light and the reception at our end. The image we see is what the object looked like when it emitted that light. If we look at an object 186,000 miles away, we see it as it appeared one second earlier; at 1,860,000 miles away, we see it as it appeared ten seconds earlier; and so on. For extremely distant objects, we see them as they were millions or billions of years in the past.

Now the second curiosity. Since the late 1920s we have known that the universe is expanding, and that as it does so it is thinning out and cooling. By measuring the current rate of expansion, we can make good estimates of the moment in the past when the expansion began—the Big Bang—which was about 13.7 billion years ago, a time when no planets or stars or galaxies existed and the entire universe consisted of a fantastically dense nugget of pure energy. No matter how big our telescopes, we cannot see beyond the distance light has traveled since the Big Bang. Farther than that, and there simply hasn't been enough time since the birth of the universe for light to get from there to here. This giant sphere, the maximum distance we can see, is only the *observable* universe. But the universe could extend far beyond that.

In his office in Santa Cruz, Garth Illingworth and his colleagues have mapped out and measured the cosmos to the edge of the observable universe. They have reached out almost as far as the laws of physics allow. All that exists in the knowable universe—oceans and sky; planets and stars; pulsars, quasars, and dark matter; distant galaxies and clusters of galaxies; and great clouds of star-forming gas—has been gathered within the cosmic sensorium gauged and observed by human beings.

"Every once in a while," says Illingworth, "I think, By God, we are studying things that we can never physically touch. We sit on this miserable little planet in a midsize galaxy and we can characterize most of the universe. It is astonishing to me, the immensity of the situation, and how to relate to it in terms we can understand."

*

The idea of Mother Nature has been represented in every culture on Earth. But to what extent is the new universe, vastly larger than anything conceived of in the past, part of *nature*? One wonders how connected Illingworth feels to this astoundingly large cosmic terrain, to the galaxies and stars so distant that their images have taken billions of years to reach our eyes. Are the little red dots on his maps part of the same landscape that Wordsworth and Thoreau described, part of the same environment of mountains and trees, part of the same cycle of birth and death that orders our lives, part of our physical and emotional conception of the world we live in? Or are such things instead digitized abstractions, silent and untouchable, akin to us only in their (hypothesized) makeup of atoms and molecules? And to what extent are we human beings, living on a small planet orbiting one star among billions of stars, part of that same nature?

The heavenly bodies were once considered divine, made of entirely different stuff from objects on Earth. Aristotle argued that all matter was constituted from four elements: earth, fire, water, and air. A fifth element, ether, he reserved for the heavenly bodies, which he considered immortal, perfect, and indestructible. It wasn't until the birth of modern science, in the seventeenth century, that we began to understand the similarity of heaven and Earth. In 1610, using his new telescope, Galileo noted that the sun had dark patches and blemishes, suggesting that the heavenly bodies are not perfect. In 1687 Newton proposed a universal law of gravity that would apply equally to the fall of an apple from a tree and to the orbits of planets around the sun. Newton then went further, suggesting that all the laws of nature apply to phenomena in the heavens as well as on Earth. In later centuries, scientists used our understanding of terrestrial chemistry and physics to estimate how long the sun could continue shining before depleting its resources of energy; to determine the chemical composition of stars; to map out the formation of galaxies.

Yet even after Galileo and Newton, there remained another question: Were living things somehow different from rocks and water and stars? Did animate and inanimate matter differ in some fundamental way? The "vitalists" claimed that animate matter had some special essence, an intangible spirit or soul, while the "mechanists" argued that living things were elaborate machines and obeyed precisely the same laws of physics and chemistry as did inanimate material. In the late nineteenth century, two German physiologists, Adolf Eugen Fick and Max Rubner, each began testing the mechanistic hypothesis by painstakingly tabulating the energies required for muscle contraction, body heat, and other physical activities and comparing these energies against the chemical energy stored in food. Each gram of fat, carbohydrate, and protein had its energy equivalent. Rubner concluded that the amount of energy used by a living creature was exactly equal to the energy it consumed in its food. Living things were to be viewed as complex arrangements of biological pulleys and levers, electric currents, and chemical impulses. Our bodies are made of the same atoms and molecules as stones, water, and air.

And yet many had a lingering feeling that human beings were somehow separate from the rest of nature. Such a view is nowhere better illustrated than in the painting *Tallulah Falls* (1841), by George Cooke, an artist associated with the Hudson River school. Although this group of painters celebrated nature, they also believed that human beings were set apart from the natural world. Cooke's painting depicts tiny human figures standing on a small promontory above a deep canyon. The people are dwarfed by tree-covered mountains, massive rocky ledges, and a waterfall pouring down to the canyon below. Not only insignificant in size compared with their surroundings, the human beings are mere witnesses to a scene they are not part of and never could be. Just a few years earlier, Ralph Waldo Emerson had published his famous essay "Nature," an appreciation of the natural world that nonetheless held humans separate from nature, at the very least in the moral and spiritual domain: "Man is fallen; nature is erect."

Today, with various back-to-nature movements attempting to resist the dislocations brought about by modernity, and with our awareness of Earth's precarious environmental state ever increasing, many people feel a new sympathy with the natural world on this planet. But the gargantuan cosmos beyond remains remote. We might understand at some level that those tiny points of light in the night sky are similar to our sun, made of atoms identical to those in our bodies, and that the cavern of outer space extends from our galaxy of stars to other galaxies of stars, to distances that would take light billions of years to traverse. We might understand these discoveries in intellectual terms, but they are baffling abstractions, even disturbing, like the notion that each of us once was the size of a dot, without mind or thought. Science has vastly expanded the scale of our cosmos, but our emotional reality is still limited by what we can touch with our bodies in the time span of our lives. George Berkeley, the eighteenth-century Irish philosopher, argued that the entire cosmos is a construct of our minds, that there is no material reality outside our thoughts. As a scientist, I cannot accept that belief. At the emotional and psychological level, however, I can have some sympathy with Berkeley's views. Modern science has revealed a world as far removed from our bodies as colors are from the blind.

Very recent scientific findings have added yet another dimension to the question of our place in the cosmos. For the first time in the history of science, we are able to make plausible estimates of the rate of occurrence of life in the universe. In March 2009, NASA launched a spacecraft called *Kepler,* whose mission was to search for planets orbiting in the "habitable zone" of other stars. The habitable zone is the region in which a planet's surface temperature is not so cold as to freeze water and not so hot as to boil it. For many reasons, biologists and chemists believe that liquid water is required for the emergence of life, even if that life may be very different from life on Earth. Dozens of candidates for such planets have been found, and we can make a rough preliminary calculation that something like 3 percent of all stars are accompanied by a potentially life-sustaining planet. The totality of living matter on Earth—humans and animals, plants, bacteria, and pond scum —makes up 0.00000001 percent of the mass of the planet. Combining this figure with the results from the *Kepler* mission, and assuming that all potentially life-sustaining planets do indeed have life, we can estimate that the fraction of stuff in the visible universe that exists in living form is something like 0.000000000000001 percent, or one-millionth of one-billionth of 1 percent. If some cosmic intelligence created the universe, life would seem to have been only an afterthought. And if life emerges by random processes, vast amounts of lifeless material are needed for each particle of life. Such numbers cannot help but bear upon the question of our significance in the universe.

Decades ago, when I was sailing with my wife in the Aegean Sea, in the midst of unending water and sky, I had a slight inkling of

infinity. It was a sensation I had not experienced before, accompanied by feelings of awe, fear, sublimity, disorientation, alienation,
and disbelief. I set a course for 255 degrees, trusting in my com-
pass—a tiny disk of painted numbers with a sliver of rotating metal
—and hoped for the best. In a few hours, as if by magic, a pale
ocher smidgen of land appeared dead ahead, a thing that drew
closer and closer, a place with houses and beds and other human
beings.

DAVID QUAMMEN

Out of the Wild

FROM *Popular Science*

IN JUNE 2008, a Dutch woman named Astrid Joosten left the
Netherlands with her husband for an adventure vacation in
Uganda. It wasn't their first trip to Africa, but it would be more
consequential than the others.

At home in Noord-Brabant, Joosten, forty-one, worked as a business analyst for an electrical company. Both she and her spouse, a
financial manager, enjoyed escaping Europe on annual getaways.
In 2002 they had flown to Johannesburg and, stepping off the airplane, felt love for Africa at first sight. On later trips they visited
Mozambique, Zambia, and Mali. The journey to Uganda in 2008,
booked through an adventure-travel outfitter, would allow them to
see mountain gorillas in the southwestern highlands of the country as well as some other wildlife and cultures. They worked their
way south toward Bwindi Impenetrable Forest, where the gorillas
reside. On one day, the operators offered a side trip, an option,
to a place called the Maramagambo Forest, where the chief attrac-
tion was a site known as Python Cave. African rock pythons lived
there, languid and content, grown large on a diet of bats.

Joosten's husband, later her widower, is a fair-skinned man
named Jaap Taal, a calm fellow with a shaved head and dark,
roundish glasses. Most of the other travelers didn't fancy this Python Cave offering, he told me in a subsequent interview. "But
Astrid and I always said, 'Maybe you come here only once in your
life, and you have to do everything you can.'" They rode to Maramagambo Forest and then walked a mile or so, gradually ascend-

ing, to a small pond. Nearby, half concealed by moss and other greenery, like a crocodile's eye barely surfaced, was a low, dark opening. Joosten and Taal, with their guide and one other client, climbed down into the cave.

The footing was bad: rocky, uneven, and slick. The smell was bad too: fruity and sour. Think of a dreary barroom, closed and empty, with beer on the floor at three A.M. The cave seemed to have been carved by a creek, or at least to have channeled its waters, and part of the overhead rock had collapsed, leaving a floor of boulders and coarse rubble, a moonscape, coated with guano like a heavy layer of vanilla icing. It served as a major roosting site for the Egyptian fruit bat (*Rousettus aegyptiacus*), a crow-size chiropteran that's widespread and relatively abundant in Africa and the Middle East. The cave's ceiling was thick with them—many thousands, agitated and chittering at the presence of human intruders, shifting position, some dropping free to fly and then settling again. Joosten and Taal kept their heads low and watched their step, trying not to slip, ready to put a hand down if needed. "I think that's how Astrid got infected," Taal told me. "I think she put her hand on a piece of rock, which contained droppings of a bat, which are infected. And so she had it on her hand." Maybe she touched her face an hour later, or put a piece of candy in her mouth, "and that's how I think the infection got in her."

No one had warned Joosten and Taal about the potential hazards of an African bat cave. They knew nothing of a virus called Marburg (though they had heard of Ebola). They stayed in the cave only about ten minutes. They saw a python, large and torpid. Then they left, continued their Uganda vacation, visited the mountain gorillas, took a boat trip, and flew back to Amsterdam. Thirteen days after the cave visit, home in Noord-Brabant, Joosten fell sick.

At first it seemed no worse than the flu. Then her temperature climbed higher and higher. After a few days, she began suffering organ failure. Her doctors, knowing of her recent time in Africa, suspected Lassa virus or maybe Marburg. "Marburg?" said Taal. "What's that?" Joosten's brother looked it up on Wikipedia and told him: "Marburg virus: it kills, could be big trouble." In fact, it's a filovirus, the closest relative to the ebolaviruses (of which there are five species, including the most infamous, Ebola). Marburg was first discovered in 1967, when a group of African monkeys, imported to Marburg an der Lahn, in western Germany, for medical research uses, passed a nasty new virus to laboratory workers. Five people died. In the decades since, it has also struck hundreds of Africans, with a case fatality rate of up to 90 percent.

The doctors moved Joosten to a hospital in Leiden, where she could get better care and be isolated from other patients. There she developed a rash and conjunctivitis; she hemorrhaged. She was put into an induced coma, a move dictated by the need to dose her more aggressively with antiviral medicine. Before she lost consciousness, though not long before, Taal went back into the isolation room, kissed his wife, and said to her, "Well, we'll see you in a few days." Blood samples, sent to a lab in Hamburg, confirmed the diagnosis: Marburg. She worsened. As her organs shut down, she lacked for oxygen to the brain, she suffered cerebral edema, and before long Joosten was declared brain-dead. "They kept her alive for a few more hours, until the family arrived," Taal told me. "Then they pulled the plug out, and she died within a few minutes."

A horse dies mysteriously in Australia, and people around it fall sick. A chimpanzee carcass in central Africa passes Ebola to the villagers who scavenge and eat it. A palm civet, served at a Wild Flavors restaurant in southern China, infects one diner with a new ailment, which spreads to Hong Kong, Toronto, Hanoi, and Singapore, eventually to be known as SARS. These cases and others, equally spooky, represent not isolated events but a pattern, a trend: the emergence of new human diseases from wildlife.

The experts call such diseases zoonoses, meaning animal infections that spill into people. About 60 percent of human infectious diseases are zoonoses. For the most part, they result from infection by one of six types of pathogens: viruses, bacteria, fungi, protists, prions, and worms. The most troublesome are viruses. They are abundant, adaptable, not subject to antibiotics, and only sometimes deterred by antiviral drugs. Within the viral category is one particularly worrisome subgroup, RNA viruses. AIDS is caused by a zoonotic RNA virus. So was the 1918 influenza, which killed 50 million people. Ebola is an RNA virus, which emerged in Uganda this summer after four years of relative quiescence. Marburg, Lassa, West Nile, Nipah, dengue, rabies, yellow fever virus, and the SARS bug are too.

Over the last half dozen years, I have asked eminent disease scientists and public-health officials, including some of the world's experts on Ebola, on SARS, on bat-borne viruses, on HIV-1 and HIV-2, and on viral evolution, the same two-part question: (1) Will a new disease emerge in the near future, sufficiently virulent and transmissible to cause a pandemic capable of killing tens of millions of people? (2) If so, what does it look like and where does it come from? Their answers to the first part have ranged from maybe to probably. Their answers to the second have focused on zoonoses, particularly RNA viruses. The prospect of a new viral pandemic, for these sober professionals, looms large. They talk about it; they think about it; they make contingency plans against it: the Next Big One. They say it might happen anytime.

To understand what killed Astrid Joosten, and to see her case within the context of the Next Big One, you need to understand how viruses evolve. Edward C. Holmes is one of the world's leading experts in viral evolution. He sits in a bare office at the Center for Infectious Disease Dynamics, which is part of Pennsylvania State University, and discerns patterns of viral change by scrutinizing sequences of genetic code. That is, he looks at long runs of the five letters (A, C, T, G, and U) that represent nucleotide bases in a DNA or RNA molecule, strung out in unpronounceable streaks as though typed by a manic chimpanzee. Holmes's office is tidy and comfortable, furnished with a desk, a table, and several chairs. There are few bookshelves, few books, few files or papers. A thinker's room. On the desk is a computer with a large monitor. That's how it all looked when I visited, anyway.

Above the computer was a poster celebrating "the Virosphere," meaning the totality of viral diversity on Earth. Beside that was another poster, showing Homer Simpson as a character in Edward Hopper's famous painting *Nighthawks*. Homer is seated at the diner counter with a plate of doughnuts before him.

Holmes is an Englishman, transplanted to central Pennsylvania from London and Cambridge. His eyes bug out slightly when he discusses a crucial fact or an edgy idea, because good facts and ideas impassion him. His head is round and, where not already bald, shaved austerely. He wears wiry glasses with a thick metal brow, and while he looks a bit severe, Holmes is anything but.

He's lively and humorous, a generous soul who loves conversation about what matters: viruses. Everyone calls him Eddie.

"Most emerging pathogens are RNA viruses," he told me as we sat beneath the two posters. RNA as opposed to DNA viruses, he meant, or to bacteria or to any other type of pathogen. To say that Eddie Holmes wrote the book on this subject wouldn't be metaphorical. It's titled *The Evolution and Emergence of RNA Viruses,* published by Oxford University Press in 2009, and that's what had brought me to his door. Now he was summarizing some of the highlights.

There are an *awful* lot of RNA viruses, he said, which might seem to raise the odds that many would come after humans. RNA viruses in the oceans, in the soil, in the forests, and in the cities; RNA viruses infecting bacteria, fungi, plants, and animals. It's possible that every cellular species of life on the planet supports at least one RNA virus, though we don't know for sure because we've just begun looking. A glance at his Virosphere poster, which portrayed the universe of known viruses as a brightly colored pizza, was enough to support that point. It showed RNA viruses accounting for at least half the slices. But they're not merely common, Eddie said. They're also highly evolvable. They're protean. They adapt quickly.

Two reasons for that, he explained. It's not just the high mutation rates but also the fact that their population sizes are huge. "Those two things put together mean you'll produce more adaptive change," he said.

RNA viruses replicate quickly, generating big populations of viral particles within each host. Stated another way, they tend to produce acute infections, severe for a short time and then gone. Either they soon disappear or they kill you. Eddie called it "this kind of boom-bust thing." Acute infection also means lots of viral shedding—by way of sneezing or coughing or vomiting or bleeding or diarrhea—which facilitates transmission to other victims. Such viruses try to outrace the immune system of each host, taking what they need and moving onward quickly, before a body's defenses can defeat them. (The HIVs are an exception, using a slower strategy.) Their fast replication and high rates of mutation supply them with lots of genetic variation. Once an RNA virus has landed in another host—sometimes even another *species* of host

—that abundant variation serves it well, giving it many chances to adapt to the new circumstances, whatever those circumstances might be.

Most DNA viruses embody the opposite extremes. Their mutation rates are low and their population sizes can be small. Their strategies of self-perpetuation "tend to go for this persistence route," Eddie said. Persistence and stealth. They lurk; they wait. They hide from the immune system rather than trying to outrun it. They go dormant and linger within certain cells, replicating little or not at all, sometimes for many years. I knew he was talking about things like varicella zoster, a classic DNA virus that begins its infection of humans as chickenpox and can recrudesce, decades later, as shingles. The downside for DNA viruses, he said, is that they can't adapt so readily to a new species of host. They're just too stable. Hidebound. Faithful to what has worked in the past.

The stability of DNA viruses derives from the structure of the genetic molecule and how it replicates: it uses the enzyme DNA polymerase to assemble and proofread each new strand. The enzyme employed by RNA viruses, on the other hand, is "error-prone," according to Eddie. "It's just a really crappy polymerase," which doesn't proofread, backtrack, or correct erroneous placement of those RNA nucleotide bases, A, C, G, and U. Why not? Because the genomes of RNA viruses are tiny, ranging from about 3,000 nucleotides to about 30,000, which is much less than what most DNA viruses carry. "It takes more nucleotides," Eddie said —a larger genome, more information—"to make a new enzyme that works." One that works as neatly as DNA polymerase does, he meant.

And why are RNA genomes so small? Because their self-replication is so fraught with inaccuracies that if given more information to replicate, they would accumulate more errors and cease to function at all. It's sort of a chicken-and-egg problem. RNA viruses are limited to small genomes because their mutation rates are so high, and their mutation rates are so high because they're limited to small genomes. In fact, there's a fancy name for that bind: Eigen's paradox. Manfred Eigen is a German chemist, a Nobel laureate, who has studied the evolution of large, self-replicating molecules. His paradox describes a size limit for such molecules, beyond which their mutation rate gives them too many errors and they cease to replicate. They die out. RNA viruses, thus constrained, compensate for their error-prone replication by producing huge populations and achieving transmission early and often. They can't break through Eigen's paradox, it seems, but they can scoot around it, making a virtue of their instability. Their copying errors deliver lots of variation, and variation allows them to evolve fast.

"DNA viruses can make much bigger genomes," Eddie said. Unlike the RNAs, they're not limited by Eigen's paradox. They can even capture and incorporate genes from the host, which helps them confuse a host's immune response. They can reside in a body for longer stretches of time, content to get themselves passed along by slower modes of transmission, such as sexual and mother-to-child. "RNA viruses can't do that." They face a different set of limits and options. Their mutation rates can't be lowered. Their genomes can't be enlarged. "They're kind of stuck."

What do you do if you're a virus that's stuck, with no long-term security, no time to waste, nothing to lose, and a high capacity for adapting to new circumstances? By now we had worked our way around to the point that interested me most. "They jump species a lot," Eddie said.

Whence do they jump? From one species of primate to another, from one rodent to another, from a prey animal into a predator, and so on. Such leaps probably occur often in the quiet isolation of forests and other wild habitats, and usually they go undetected by science. But sometimes the leap is from a nonhuman critter into a human. Then we notice.

The kind of animal that harbors a given virus is known as its reservoir host. Could be a monkey, a bat, maybe a rat. Within its reservoir host the virus lives quietly, in a sort of long-term truce, causing no obvious symptoms. Passage from one kind of host to another is called spillover. In the new host, the old truce doesn't apply. The virus may turn aggressive and virulent. If the new host is human, you've got a newly emerged zoonotic disease.

Spillover to humans, as Eddie Holmes noted, occurs more often among RNA viruses than other bugs. It brings creatures such as Lassa (first recorded in 1969), Ebola (1976), HIV-1 (inferred in 1981, isolated in 1983), HIV-2 (1986), Sin Nombre (the infamous American hantavirus, 1993), Hendra (1994), avian flu (1997), Nipah (1998), West Nile (1999), SARS (2003), and swine flu (2009) into people's lives. Marburg is just another of the leaping threats, rare but dramatic in its impact on humans. Why are these spillovers happening ever more frequently, in what seems a drumbeat of bad news?

To put the matter in its starkest form: human-caused ecological pressures and disruptions are bringing animal pathogens ever more into contact with human populations, while human technology and behavior are spreading those pathogens ever more widely and quickly. In other words, outbreaks of new zoonotic diseases, as well as the recurrence and spread of old ones, reflect things that we're *doing,* rather than just being things that are *happening to* us.

We have increased our human population to the level of 7 billion and beyond. We are well on our way toward 9 billion before our growth trend is likely to flatten. We live at high densities in many cities. We have penetrated, and we continue to penetrate, the last great forests and other wild ecosystems of the planet, disrupting the physical structures and the ecological communities of such places. We cut our way through the Congo. We cut our way through the Amazon. We cut our way through Borneo. We cut our way through Madagascar. We cut our way through New Guinea and northeastern Australia. We shake the trees, figuratively and literally, and things fall out. We kill and butcher and eat many of the wild animals found there. We settle in those places, creating villages, work camps, towns, extractive industries, new cities. We bring in our domesticated animals, replacing the wild herbivores with livestock. We multiply our livestock as we've multiplied ourselves, establishing huge factory-scale operations that contain thousands of cattle, pigs, chickens, ducks, sheep, and goats. We export and import livestock, fed and fattened with prophylactic doses of antibiotics and other drugs, across great distances and at high speeds. We export and import wild animals as exotic pets. We export and import animal skins, contraband bushmeat, and plants, some of which carry hidden microbial passengers. We travel, moving between cities and continents even more quickly than our transported livestock. We visit monkey temples in Asia, live markets in India, picturesque villages in South America, dusty archaeological sites in New Mexico, dairy towns in the Netherlands, bat caves in East Africa, racetracks in Australia—breathing the air, feeding the animals, touching things, shaking hands with the locals—and then we jump on our planes and fly home. We provide an irresistible opportunity for enterprising microbes by the ubiquity and sheer volume and mass of our human bodies.

Everything just mentioned falls under this rubric: the ecology and evolutionary biology of zoonotic diseases. Ecological circumstance provides opportunity for spillover. Evolution seizes opportunity, explores possibilities, and helps convert spillovers to pandemics. But "ecology" and "evolutionary biology" sound like science, not medicine or public health. If zoonoses from wildlife represent such a significant threat to global security, then what's to be done? Learn more. RNA viruses are everywhere, as Eddie Holmes has warned, and science has identified only a fraction of them. Fewer still have been traced to their reservoir hosts, isolated from the wild, grown in the lab, and systematically studied. Until those steps have been achieved, the viruses in question can't be battled with vaccines and treatments. This is where the field and laboratory scientists—veterinary ecologists, epidemiologists, molecular phylogeneticists, lab virologists—come in. If we're going to understand how zoonoses operate, we need to find these bugs in the world, grow them in cell cultures the old-fashioned way, look at them in the flesh, sequence their genomes, and place them within their family trees. It's happening, in laboratories and at field sites all over the world, but it's no simple task.

Astrid Joosten wasn't the only person in recent years to die of Marburg. In 2007, a year before her visit to Uganda, a small outbreak occurred among miners in roughly the same area. Just four men were affected, of whom one died. All of them worked at a site called Kitaka Cave, in the southwestern corner of Uganda.

Soon after the news of the affliction got out, in August 2007, an international response team converged on Uganda to assist and collaborate with the Ugandan Ministry of Health. The group included scientists from the Centers for Disease Control and Prevention (CDC) in Atlanta, the National Institute for Communicable Diseases (NICD) in South Africa, and the World Health Organization (WHO) in Geneva. From the CDC there was Pierre Rollin, an expert on the filoviruses and their clinical impacts. Along with him from Atlanta had come Jonathan Towner, Brian Amman, and Serena Carroll. Pierre Formenty had arrived from the WHO; Bob Swanepoel and Alan Kemp of the NICD had flown up from Johannesburg. All of them possessed extensive experience with Ebola and Marburg, gained variously through outbreak responses, lab research, and field studies.

The cave served as the roosting site for about 100,000 individuals of the Egyptian fruit bat, then a prime suspect as a reservoir for Marburg. The team members, wearing Tyvek suits, rubber boots, goggles, respirators, gloves, and helmets, had been shown to the shaft by miners, who as usual were clad only in shorts, T-shirts, and sandals. Guano covered the ground. The miners clapped their hands to scatter low-hanging bats as they went. The bats, panicked, came streaming out. These were sizable animals, each with a two-foot wingspan, not quite so large and hefty as some fruit bats but still daunting, especially with thousands swooshing at you in a narrow tunnel. Before he knew it, Amman had been conked in the face by a bat and taken a cut over one eyebrow. Towner got hit too. Fruit bats have long, sharp thumbnails. Later, because of the cut, Amman would get a postexposure shot against rabies, though Marburg was a more immediate concern. "Yeah," he thought, "this could be a really good place for transmission."

The cave had several shafts. The main shaft was about eight feet high. Because of all the mining activity, many of the bats had shifted their roosting preference "and went over to what we called the cobra shaft," Amman later told me. The shaft was so called because, he said, "there was a black forest cobra in there."

Or maybe a couple. It was a good dark habitat for a snake, with water and plenty of bats to eat. The miners showed Amman and Towner into the cave and led them to a chamber containing a body of brown, tepid water. Then the local fellows cleared out, leaving the scientists to explore on their own. They dropped down beside the brown lake and found that the chamber branched into three shafts, each of which seemed blocked by standing water. Peering into those shafts, they could see many more bats. The humidity was high and the temperature maybe 10 or 15 degrees hotter than outside. Their goggles fogged up. Their respirators became soggy and wouldn't pass much oxygen. They were panting and sweating, zipped into their Tyvek suits, which felt like wearing a trash bag, and by now they were becoming "a little loopy," Amman recalled. "We had to get out and cool off." It was only their first underground excursion at Kitaka. They would make several.

On a later day, the team investigated a grim, remote chamber they dubbed the Cage. It was where one of the four infected miners had been working just before he got sick. This time Amman, Formenty, and Alan Kemp of the NICD went to the far recesses of the cave. The Cage itself could be entered only by crawling through a low gap at the base of a wall—like sliding under a garage door that hadn't quite closed. Amman is a large man, six-foot-three and 220 pounds, and for him the gap was a tight squeeze; his helmet got stuck, and he had to pull it through separately. "You come out into this sort of blind room," he said, "and the first thing you see is just hundreds of these dead bats."

They were Egyptian fruit bats, the creature of interest, left in various stages of mummification and rot. Piles of dead and liquescent bats seemed a bad sign, potentially invalidating the hypothesis that *Rousettus aegyptiacus* might be a reservoir host of Marburg. If these bats *had* died of Marburg, suspicion would shift elsewhere —to another bat or maybe a rodent or a tick or a spider. Those other suspects might have to be investigated. Ticks, for instance: there were plenty of them in crevices near the bat roosts, waiting for a chance to drink some blood.

The men went to work, collecting. They stuffed dead bats into bags. They caught a few live bats and bagged them too. Then, back down on their bellies, they squeezed out through the low gap. "It was really unnerving," Amman told me. "I'd probably never do it again. One little accident, a big rock rolls in the way, and that's it. You're trapped. Uganda is not famous for its mine rescue teams."

By the end of this field trip, the scientists had collected about 800 bats. They dissected them and took samples of blood and tissue. Those samples went back to Atlanta, where Towner participated in the laboratory efforts to find traces of Marburg virus. One year later came a paper, authored by Towner, Amman, Rollin, and their WHO and NICD colleagues, announcing some important results. Not only did the team detect antibodies against Marburg and fragments of Marburg RNA, but they also did something more difficult and compelling. They found live virus.

Working in one of the CDC's Biosafety Level 4 units (the highest level of containment security for pathogens), Towner and his coworkers had isolated viable, replicating Marburg virus from five different bats. Furthermore, the five strains of virus were genetically diverse, suggesting an extended history of viral presence and evolution within Egyptian fruit bats. That data, plus the fragmentary RNA, constituted strong evidence that the bat is a reservoir—if not *the* reservoir—of Marburg virus. The virus is definitely there, infecting about 5 percent of the bat population at a given time. Of the estimated 100,000 bats at Kitaka, therefore, the team could say that about 5,000 Marburg-infected bats flew out of the cave every night.

An interesting thought: 5,000 infected bats passing overhead. Where were they going? How far to the fruiting trees? Whose livestock or little gardens got shat upon as they went? The breadth of possible transmission is incalculable. And the Kitaka aggregation, Towner and his coauthors added, "is only one of many such cave populations throughout Africa."

The dangers presented by zoonoses are real and severe, but the degree of uncertainty is also high. There's not a hope in hell, for instance, as a great flu expert told me, of predicting the nature and timing of the next influenza pandemic. Too many factors vary randomly, or almost randomly, in that system. Prediction, in general, so far as all these diseases are concerned, is a tenuous proposition, more likely to yield false confidence than actionable intelligence.

But the difficulty of predicting precisely doesn't oblige us to remain blind, unprepared, and fatalistic about emerging and reemerging zoonotic diseases. The practical alternative to soothsaying, as one expert put it, is "improving the scientific basis to improve readiness." By "the scientific basis" he meant the understanding of which virus groups to watch, the field capabilities to detect spillovers in remote places before they become regional outbreaks, the organizational capacities to control outbreaks before they become pandemics, plus the laboratory tools and skills to recognize known viruses speedily, to characterize new viruses almost as fast, and to create vaccines and therapies without much delay. If we can't predict a forthcoming influenza pandemic or any other newly emergent virus, we can at least be vigilant; we can be well prepared and quick to respond; we can be ingenious and scientifically sophisticated in the forms of our response.

To a considerable degree, such things are already being done. Ambitious networks and programs have been created by the WHO, the CDC, and other national and international agencies to address the danger of emerging zoonotic diseases. Because of concern over the potential of "bioterrorism," even the U.S. Department of Homeland Security and the Defense Advanced Research Projects Agency (DARPA, whose motto is "Creating & Preventing Strategic Surprise") of the U.S. Department of Defense have their hands in the mix. These efforts carry names and acronyms such as the Global Outbreak Alert and Response Network (GOARN, of the WHO), Prophecy (of DARPA), the Emerging Pandemic Threats program (EPT, of USAID), and the Special Pathogens Branch (SPB, of the CDC), all of which sound like programmatic boilerplate but which harbor some dedicated people working in field sites where spillovers happen and secure labs where new pathogens can be quickly studied. Private organizations such as EcoHealth Alliance (led by a former parasitologist named Peter Daszak) have also tackled the problem. There is an intriguing organization called Global Viral (GV), created by a scientist named Nathan Wolfe and financed in part by Google. GV gathers blood samples on small patches of filter paper from bushmeat hunters and other people across tropical Africa and Asia and screens those samples for new viruses, in a systematic effort to detect spillovers and stop the next pandemic before it begins to spread. At the Mailman School of Public Health, part of Columbia University, researchers in Ian Lipkin's laboratory are developing new molecular diagnostic tools. Lipkin, trained as a physician as well as a molecular biologist, calls his métier "pathogen discovery" and uses techniques such as high-throughput sequencing (which can sequence thousands of DNA samples quickly and cheaply), MassTag-PCR (identifying amplified genome segments by mass spectrometry), and the GreeneChip diagnostic system, which can simultaneously screen for thousands of different pathogens. When a field biologist takes serum from flying foxes in Bangladesh or bleeds little bats in southern China, some of those samples go straight to Lipkin.

These scientists are on alert. They are our sentries. They watch the boundaries across which pathogens spill. When the next novel virus makes its way from a chimpanzee, a bat, a mouse, a duck, or a macaque into a human, and maybe from that human into another human, and thereupon begins causing a small cluster of lethal illnesses, they will see it—we hope they will, anyway—and raise the alarm.

*

During the early twentieth century, disease scientists from the Rockefeller Foundation and other institutions conceived the ambitious goal of eradicating some infectious diseases entirely. They tried hard with yellow fever, spending millions of dollars and many years of effort, and failed. They tried with malaria and failed. They tried later with smallpox and succeeded. Why? The differences among those three diseases are many and complex, but probably the most crucial one is that smallpox resided neither in a reservoir host nor in a vector, such as a mosquito or tick. Its ecology was simple. It existed in humans—in humans only—and was therefore much easier to eradicate. The campaign to eradicate polio, begun in 1998 by the WHO and other institutions, is a realistic effort for the same reason: polio isn't zoonotic. Eradicating a zoonotic disease, whether a directly transmitted one like Ebola or an insect-vectored one such as yellow fever, is much more complicated. Do you exterminate the pathogen by exterminating the species of bat or primate or mosquito in which it resides? Not easily, you don't, and not without raising an outcry. The notion of eradicating chimpanzees as a step toward preventing the future spillover of another HIV would provoke a deep and bitter discussion, to put it mildly.

That's the salubrious thing about zoonotic diseases: they remind us, as Saint Francis did, that we humans are inseparable from the natural world. In fact, there *is* no "natural world"; it's a bad and artificial phrase. There is only the world. Humankind is part of that world, as are the ebolaviruses, as are the influenzas and the HIVs, as are Marburg and Nipah and SARS, as are chimpanzees and palm civets and Egyptian fruit bats, as is the next murderous virus—the one we haven't yet detected. And while humans don't evolve nearly as fast and as variously as an RNA virus does, we may —let me repeat that word, *may*—be able to keep such threats at bay, fighting them off, forestalling the more cataclysmic of the dire scenarios they present, for one reason: at our best, we're smarter than they are.

OLIVER SACKS

Altered States

FROM *The New Yorker*

To LIVE ON a day-to-day basis is insufficient for human beings; we need to transcend, transport, escape; we need meaning, understanding, and explanation; we need to see overall patterns in our lives. We need hope, the sense of a future. And we need freedom (or, at least, the illusion of freedom) to get beyond ourselves, whether with telescopes and microscopes and our ever-burgeoning technology, or in states of mind that allow us to travel to other worlds, to rise above our immediate surroundings.

We may seek, too, a relaxing of inhibitions that makes it easier to bond with each other or transports that make our consciousness of time and mortality easier to bear. We seek a holiday from our inner and outer restrictions, a more intense sense of the here and now, the beauty and value of the world we live in.

Many of us find Wordsworthian "intimations of immortality" in nature, art, creative thinking, or religion; some people can reach transcendent states through meditation or similar trance-inducing techniques, or through prayer and spiritual exercises. But drugs offer a shortcut; they promise transcendence on demand. These shortcuts are possible because certain chemicals can directly stimulate many complex brain functions.

Every culture has found such chemical means of transcendence, and at some point the use of such intoxicants becomes institutionalized at a magical or sacramental level. The sacramental use of psychoactive plant substances has a long history and continues to the present day in various shamanic and religious rites around the world.

At a humbler level, drugs are used not so much to illuminate or expand or concentrate the mind but for the sense of pleasure and euphoria they can provide. Even the pioneer Mormons, forbidden to use tea or coffee, on their long march to Utah found by the roadside a simple herb, Mormon tea, whose infusions refreshed and stimulated the weary pilgrims. This was ephedra, which contains ephedrine, chemically and pharmacologically akin to the amphetamines.

Many people experiment with drugs, hallucinogenic and otherwise, in their teenage or college years. I did not try them until I was thirty and a neurology resident. This long virginity was not due to lack of interest. I had read the great classics—De Quincey's *Confessions of an English Opium Eater* and Baudelaire's *Artificial Paradises*— at school. I read about the French writer Théophile Gautier, who in 1845 paid a visit to the recently founded Club des Hashischins, in a quiet corner of the Île Saint-Louis. Hashish, in the form of a greenish paste, had recently been introduced from Algeria and was all the rage in Paris. At the salon, Gautier consumed a substantial piece of hash. At first he felt nothing out of the ordinary, but soon, he wrote, "everything seemed larger, richer, more splendid," and then more specific changes occurred:

> An enigmatic personage suddenly appeared before me . . . His nose was bent like the beak of a bird, his green eyes, which he wiped frequently with a large handkerchief, were encircled with three brown rings, and caught in the knot of a high white starched collar was a visiting card which read: *Daucus-Carota, du Pot d'or* . . . Little by little the salon was filled with extraordinary figures, such as are found only in the etchings of Callot or the aquatints of Goya; a pêle-mêle of rags and tatters, bestial and human shapes.

By the 1890s, Westerners were also beginning to sample mescal, or peyote, previously used only as a sacrament in certain Native American traditions. As a freshman at Oxford, free to roam the shelves of the Radcliffe Science Library, I read the first published accounts of mescal intoxication, including those of Havelock Ellis and Silas Weir Mitchell. They were primarily medical men, not just literary ones, and this seemed to lend an extra weight and credibility to their descriptions. I was captivated by Mitchell's dry tone and his nonchalance about taking what was then an unknown drug with unknown effects.

At one point, Mitchell wrote in an 1896 article for the *British Medical Journal*, he took a fair portion of an extract made from mescal buttons and followed it up with an additional dose. Although he noted that his face was flushed, his pupils were dilated, and he had "a tendency to talk, and now and then . . . misplaced a word," he nevertheless went out on house calls and saw several patients. Afterward, following three further doses, he lay down quietly in a dark room, whereupon he experienced "an enchanted two hours," full of chromatic effects:

> Delicate floating films of colour—usually delightful neutral purples and pinks. These came and went—now here, now there. Then an abrupt rush of countless points of white light swept across the field of view, as if the unseen millions of the Milky Way were to flow a sparkling river before the eye.

Unlike Mitchell, who had focused on colored, geometric hallucinations, which he compared in part to those of migraine, Aldous Huxley, writing of mescaline in the 1950s, focused on the transfiguration of the visual world, its investment with luminous, divine beauty and significance. He compared such drug experiences to those of great visionaries and artists, though also to the psychotic experiences of some schizophrenics. Both genius and madness, Huxley hinted, lay in these extreme states of mind—a thought not so different from those expressed by De Quincey, Coleridge, and Baudelaire in relation to their own ambiguous experiences with opium and hashish (and explored at length in Moreau's 1845 book *Hashish and Mental Illness*). I read Huxley's *The Doors of Perception* and *Heaven and Hell* when they came out in the 1950s, and I was especially excited by his speaking of the geography of the imagination and its ultimate realm—the "antipodes of the mind."

I had done a great deal of reading but had no experiences of my own with such drugs until 1953, when my childhood friend Eric Korn came up to Oxford. We read excitedly about Albert Hofmann's discovery of LSD, and we ordered 50 micrograms of it from the manufacturer in Switzerland (it was still legal in the midfifties). Solemnly, even sacramentally, we divided it and took 25 micrograms each—not knowing what splendors or horrors awaited us—but, sadly, it had absolutely no effect on either of us. (We should have ordered 500 micrograms, not 50.)

By the time I qualified as a doctor at the end of 1958, I knew I wanted to be a neurologist, to know how the brain embodied consciousness and self and to understand its amazing powers of perception, imagery, memory, and hallucination. A new orientation was entering neurology and psychiatry at that time; it was the opening of a neurochemical age, with a glimpse of the range of chemical agents, neurotransmitters, which allowed nerve cells and different parts of the nervous system to communicate with one another. In the 1950s and '60s, discoveries were coming from all directions, though it was far from clear how they fitted together. It had been found, for instance, that the Parkinsonian brain was low in dopamine, and that giving a dopamine precursor, L-dopa, could alleviate the symptoms of Parkinson's disease; while tranquilizers, introduced in the early 1950s, could depress dopamine and cause a sort of chemical Parkinsonism. For about a century, the staple medication for Parkinsonism had been anticholinergic drugs. How did the dopamine and the acetylcholine systems interact? Why did opiates—or cannabis—have such strong effects? Did the brain have special opiate receptors and make opioids of its own? Was there a similar mechanism for cannabis receptors and cannabinoids? Why was LSD so enormously potent? Were all its effects explicable in terms of altering the serotonin in the brain? What transmitter systems governed wake-sleep cycles, and what might be the neurochemical background of dreams or hallucinations?

Starting a neurology residency in 1962, I found the atmosphere heady with such questions. Neurochemistry was plainly "in," and so—dangerously, seductively, especially in California, where I was studying—were the drugs themselves.

I started with cannabis. A friend in Topanga Canyon, where I lived at the time, offered me a joint; I took two puffs and was transfixed by what happened then. I gazed at my hand, and it seemed to fill my visual field, getting larger and larger while at the same time moving away from me. Finally, it seemed to me, I could see a hand stretched across the universe, light-years or parsecs in length. It still looked like a living, human hand, yet this cosmic hand somehow also seemed like the hand of God. My first pot experience was marked by a mixture of the neurological and the divine.

On the West Coast in the early 1960s, LSD and morning glory seeds were readily available, so I sampled those, too. "But if you want a really far-out experience," my friends on Muscle Beach told me, "try Artane." I found this surprising, for I knew that Artane, a synthetic drug allied to belladonna, was used in modest doses (two or three tablets a day) for the treatment of Parkinson's disease and that such drugs, in large quantities, could cause a delirium. (Such deliriums have long been observed with accidental ingestion of plants like deadly nightshade, thornapple, and black henbane.) But would a delirium be fun? Or informative? Would one be in a position to observe the aberrant functioning of one's brain — to appreciate its wonder? "Go on," my friends urged. "Just take twenty of them—you'll still be in partial control."

So one Sunday morning I counted out twenty pills, swallowed them with a mouthful of water, and sat down to await the effect. Would the world be transformed, newborn, as Huxley described in *The Doors of Perception,* and as I myself had experienced with mescaline and LSD? Would there be waves of delicious, voluptuous feeling? Would there be anxiety, disorganization, paranoia? I was prepared for all of these, but none of them occurred. I had a dry mouth and large pupils, and found it difficult to read, but that was all. There were no psychic effects whatever—most disappointing. I did not know exactly what I expected, but I expected something.

I was in the kitchen, putting on a kettle for tea, when I heard a knocking at my front door. It was my friends Jim and Kathy; they often dropped round on a Sunday morning. "Come in, door's open," I called out, and as they settled themselves in the living room I asked, "How do you like your eggs?" Jim liked them sunny side up, he said. Kathy preferred them over easy. We chatted away while I sizzled their ham and eggs—there were low swinging doors between the kitchen and the living room, so we could hear each other easily. Then, five minutes later, I shouted, "Everything's ready," put their ham and eggs on a tray, walked into the living room—and found it empty. No Jim, no Kathy, no sign that they had ever been there. I was so staggered I almost dropped the tray.

It had not occurred to me for an instant that Jim and Kathy's voices, their "presences," were unreal, hallucinatory. We had had a friendly, ordinary conversation, just as we usually had. Their voices were the same as always—there was no hint, until I opened the swinging doors and found the living room empty, that the whole conversation, at least their side of it, had been invented by my brain.

I was not only shocked but rather frightened, too. With LSD and other drugs, I knew what was happening. The world would look different, feel different, there would be every characteristic of a special, extreme mode of experience. But my "conversation" with Jim and Kathy had no special quality, it was entirely commonplace, with nothing to mark it as a hallucination. I thought about schizophrenics conversing with their "voices," but typically the voices of schizophrenia are mocking or accusing, not talking about ham and eggs and the weather.

"Careful, Oliver," I said to myself. "Take yourself in hand. Don't let this happen again." Sunk in thought, I slowly ate my ham and eggs (Jim and Kathy's, too) and then decided to go down to the beach, where I would see the real Jim and Kathy and all my friends and enjoy a swim and an idle afternoon.

I was pondering all this when I became conscious of a whirring noise above me. It puzzled me for a moment, and then I realized that it was a helicopter preparing to descend, and that it contained my parents, who, wanting to make a surprise visit, had flown in from London and, arriving in Los Angeles, had chartered a helicopter to bring them to Topanga Canyon. I rushed into the bathroom, had a quick shower, and put on a clean shirt and pants —the most I could do in the three or four minutes before they arrived. The throb of the engine was almost deafeningly loud, so I knew that the helicopter must have landed on the flat rock beside my house. I raced out excitedly to greet my parents—but the rock was empty, there was no helicopter in sight, and the huge pulsing noise of its engine was abruptly cut off. The silence and emptiness, the disappointment, reduced me to tears. I had been so joyful, and now there was nothing at all.

I went back into the house and put on the kettle for another cup of tea, when my attention was caught by a spider on the kitchen wall. As I drew nearer to look at it, the spider called out, "Hello!" It did not seem at all strange to me that a spider should say hello (any more than it seemed strange to Alice when the White Rabbit spoke). I said, "Hello yourself," and with this we started a conversation, mostly on rather technical matters of analytic philosophy. Perhaps this direction was suggested by the spider's opening comment: Did I think that Bertrand Russell had exploded Frege's paradox? Or perhaps it was its voice—pointed, incisive, and just like Russell's voice, which I had heard on the radio. (Decades later, I mentioned the spider's Russellian tendencies to my friend Tom Eisner, an entomologist; he nodded sagely and said, "Yes, I know the species.")

During the week, I would avoid drugs, working as a resident at UCLA's neurology department. I was amazed and moved, as I had been as a medical student in London, by the range of patients' neurological experiences, and I found that I could not comprehend these sufficiently, or come to terms with them emotionally, unless I attempted to describe or transcribe them. It was then that I wrote my first published papers and my first book. (It was never published, because I lost the manuscript.)

But on the weekends I often experimented with drugs. I recall vividly one episode in which a magical color appeared to me. I had been taught as a child that there were seven colors in the spectrum, including indigo. (Newton had chosen these, somewhat arbitrarily, by analogy with the seven notes of the musical scale.) But few people agree on what "indigo" is.

I had long wanted to see "true" indigo, and thought that drugs might be the way to do this. So one sunny Saturday in 1964 I developed a pharmacologic launch pad consisting of a base of amphetamine (for general arousal), LSD (for hallucinogenic intensity), and a touch of cannabis (for a little added delirium). About twenty minutes after taking this, I faced a white wall and exclaimed, "I want to see indigo now—now!"

And then, as if thrown by a giant paintbrush, there appeared a huge, trembling, pear-shaped blob of the purest indigo. Luminous, numinous, it filled me with rapture: it was the color of heaven, the color, I thought, that Giotto spent a lifetime trying to get but never achieved—never achieved, perhaps, because the color of heaven is not to be seen on Earth. But it existed once, I thought—it was the color of the Paleozoic sea, the color the ocean used to be. I leaned toward it in a sort of ecstasy. And then it suddenly disappeared, leaving me with an overwhelming sense of loss and sadness that it had been snatched away. But I consoled myself: yes, indigo exists, and it can be conjured up in the brain.

For months afterward, I searched for indigo. I turned over little

stones and rocks near my house. I looked at specimens of azurite
in the natural history museum—but even that was infinitely far
from the color I had seen. And then in 1965, when I had moved
to New York, I went to a concert at the Metropolitan Museum of
Art. In the first half, a Monteverdi piece was performed, and I was
transported. I had taken no drugs, but I felt a glorious river of
music, hundreds of years long, flowing from Monteverdi's mind
into my own. In this ecstatic mood, I wandered out during the
intermission and looked at the objects on display in the Egyptian
galleries—lapis lazuli amulets, jewelry, and so forth—and I was enchanted to see glints of indigo. I thought, Thank God, it really
exists!

During the second half of the concert, I got a bit bored and
restless, but I consoled myself, knowing that I could go out and
take a "sip" of indigo afterward. It would be there, waiting for me.
But when I went out to look at the gallery after the concert was
finished, I could see only blue and purple and mauve and puce
—no indigo. That was forty-seven years ago, and I have never seen
indigo again.

When a friend and colleague of my parents—Augusta Bonnard, a
psychoanalyst—came to Los Angeles for a year's sabbatical in 1964,
it was natural that we should meet. I invited her to my little house
in Topanga Canyon, and we had a genial dinner together. Over
coffee and cigarettes (Augusta was a chain smoker; I wondered
if she smoked even during analytic sessions), her tone changed,
and she said, in her gruff, smoke-thickened voice, "You need help,
Oliver. You're in trouble."

"Nonsense," I replied. "I enjoy life. I have no complaints. All is
well in work and love." Augusta let out a skeptical grunt but did
not push the matter further.

I had started taking LSD at this point, and if that was not available I would take morning glory seeds instead. (This was before
morning glory seeds were treated with pesticide, as they are now,
to prevent drug abuse.) Sunday mornings were usually my drug
time, and it must have been two or three months after meeting
Augusta that I took a hefty dose of Heavenly Blue morning glory
seeds. The seeds were jet black and of agate-like hardness, so I
pulverized them with a mortar and pestle and then mixed them
with vanilla ice cream. About twenty minutes after eating this, I felt

an intense nausea, but when it subsided I found myself in a realm of paradisiacal stillness and beauty, a realm outside time, which was rudely broken into by a taxi grinding and backfiring its way up the steep trail to my house. An elderly woman got out of the taxi, and, galvanized into action, I ran toward her, shouting, "I know who you are—you are a replica of Augusta Bonnard! You look like her, you have her posture and movements, but you are not her. I am not deceived for a moment." Augusta raised her hands to her temples and said, "Oy! This is worse than I realized." She got back into the taxi and took off without another word.

We had plenty to talk about the next time we met. My failure to recognize her, my seeing her as a "replica," she thought, was a complex form of defense, a dissociation that could only be called psychotic. I disagreed and maintained that my seeing her as a duplicate or impostor was neurological in origin, a disconnection between perception and feelings. The ability to identify (which was intact) was not accompanied by the appropriate feeling of warmth and familiarity, and it was this contradiction that led to the logical though absurd conclusion that she was a "duplicate." (This condition, which can occur in schizophrenia, but also with dementia or delirium, is known as Capgras syndrome.) Augusta said that whichever view was correct, taking mind-altering drugs every weekend, alone, and in high doses surely testified to some intense inner needs or conflicts, and that I should explore these with a therapist. In retrospect, I am sure she was right, and I began seeing an analyst a year later.

The summer of 1965 was a sort of in-between time: I had completed my residency at UCLA and had left California, but I had three months ahead of me before taking up a research fellowship in New York. This should have been a time of delicious freedom, a wonderful and needed holiday after the sixty- and sometimes eighty-hour workweeks I had had at UCLA. But I did not feel free. When I am not working, I get unmoored, have a sense of emptiness and structurelessness. Weekends were the danger times, the drug times, when I lived in California—and now an entire summer in my hometown, London, stretched before me like a three-month-long weekend.

It was during this idle, mischievous time that I descended deeper into drug-taking, no longer confining it to weekends. I tried intravenous injection, which I had never done before. My parents, both physicians, were away, and having the house to myself, I decided to explore the drug cabinet in their surgery, on the ground floor of our house, for something special to celebrate my thirty-second birthday. I had never taken morphine or any opiates before. I used a large syringe—why bother with piddling doses? And, after settling myself comfortably in bed, I drew up the contents of several vials, plunged the needle into a vein, and injected the morphine very slowly.

Within a minute or so, my attention was drawn to a sort of commotion on the sleeve of my dressing gown, which hung on the door. I gazed intently at this, and as I did so it resolved itself into a miniature but microscopically detailed battle scene. I could see silken tents of different colors, the largest of which was flying a royal pennant. There were gaily caparisoned horses, soldiers on horseback, their armor glinting in the sun, and men with longbows. I saw pipers with long silver pipes, raising these to their mouths, and then, very faintly, I heard their piping, too. I saw hundreds, thousands of men—two armies, two nations—preparing to do battle. I lost all sense of this being a spot on the sleeve of my dressing gown, or the fact that I was lying in bed, that I was in London, that it was 1965. Before shooting up the morphine, I had been reading Froissart's *Chronicles* and *Henry V,* and now these became conflated in my hallucination. I realized that I was gazing at Agincourt late in 1415 and looking down on the serried armies of England and France drawn up to do battle. And in the great pennanted tent, I knew, was Henry V himself. I had no sense that I was imagining or hallucinating any of this; what I saw was actual, real.

After a while, the scene started to fade, and I became dimly conscious, once more, that I was in London, stoned, hallucinating Agincourt on the sleeve of my dressing gown. It had been an enchanting and transporting experience, but now it was over. The drug effect was fading fast; Agincourt was hardly visible now. I glanced at my watch. I had injected the morphine at nine-thirty, and now it was ten. But I had a sense of something odd—it had been dusk when I took the morphine; it should now be darker still. But it was not. It was getting lighter, not darker, outside. It was ten, I now realized, but ten in the morning. I had been gazing, motionless, at my Agincourt for more than twelve hours. This shocked and sobered me and made me see how one could spend entire days, nights, weeks, even years of one's life in an opium stupor. I would make sure that my first opium experience was also my last.

At the end of that summer of 1965, I moved to New York to begin a postgraduate fellowship in neuropathology and neurochemistry. December 1966 was a bad time: I was finding New York difficult to adjust to after my years in California; a love affair had gone sour; my research was going badly; and I was discovering that I was not cut out to be a bench scientist. Depressed and insomniac, I was taking ever-increasing doses of chloral hydrate to get to sleep and was up to fifteen times the usual dose every night. And though I had managed to stockpile a huge amount of the drug—I raided the chemical supplies in the lab at work—this finally ran out on a bleak Tuesday a little before Christmas, and for the first time in several months I went to bed without my usual knockout dose. My sleep was poor, broken by nightmares and bizarre dreams, and upon waking I found myself excruciatingly sensitive to sounds. There were always trucks rumbling along the cobblestoned streets of the West Village; now it sounded as if they were crushing the cobblestones to powder as they passed.

Feeling a bit shaky, I did not ride my motorcycle to work, as usual, but took a train and a bus. Wednesday was brain-cutting day in the neuropathology department, and it was my turn to cut the brain into neat horizontal slices, to identify the main structures as I did so, and observe whether there were any departures from normal. I was usually pretty good at this, but today I found my hand trembling visibly, embarrassingly, and the anatomical names were slow in coming to mind.

When the session ended, I went across the road, as I often did, for a cup of coffee and a sandwich. As I was stirring the coffee, it suddenly turned green, then purple. I looked up, startled, and saw that a customer paying his bill at the cash register had a huge proboscidean head, like an elephant seal. Panic seized me; I slammed a five-dollar bill on the table and ran across the road to a bus. But all the passengers on the bus seemed to have smooth white heads like giant eggs, with huge glittering eyes like the faceted compound eyes of insects—their eyes seemed to move in sudden jerks, which increased the feeling of their fearsomeness and alienness. I realized that I was hallucinating or experiencing some bizarre perceptual disorder, that I could not stop what was happening in my brain, and that I had to maintain at least an external control and not panic or scream or become catatonic, faced by the bug-eyed monsters around me. The best way of doing this, I found, was to write, to describe the hallucination in clear, almost clinical detail, and, in so doing, become an observer, even an explorer, not a helpless victim, of the craziness inside me. I am never without pen and notebook, and now I wrote for dear life, as wave after wave of hallucination rolled over me.

Description, writing, had always been my best way of dealing with complex or frightening situations—though it had never been tested in so terrifying a situation. But it worked; by describing in my lab notebook what was going on, I managed to maintain a semblance of control, though the hallucinations continued, mutating all the while.

Somehow I got off at the right bus stop and onto the train, even though everything now was in motion, whirling vertiginously, tilting and even turning upside down. And I managed to get off at the right station in my neighborhood in Greenwich Village. As I emerged from the subway, the buildings around me were tossing and flapping from side to side, like flags blowing in a high wind. I was enormously relieved to make it back to my apartment without being attacked or arrested or killed by the rushing traffic on the way. As soon as I got back, I felt I had to contact somebody—someone who knew me well, who was both a doctor and a friend. The pediatrician Carol Burnett was the person: we had interned together in San Francisco five years earlier and had resumed a close friendship now that we were both in New York. Carol would understand; she would know what to do. I dialed her number with a now grossly tremulous hand. "Carol," I said, as soon as she picked up, "I want to say good-bye. I've gone mad, psychotic, insane. It started this morning and it's getting worse all the while."

"Oliver!" Carol said. "What have you just taken?"

"Nothing," I replied. 'That's why I'm so frightened." Carol thought for a moment, then asked, "What have you just stopped taking?"

'That's it!" I said. "I was taking a huge amount of chloral hydrate and ran out of it last night."

"Oliver, you chump! You always overdo things," Carol said. "You've got a classic case of the DTs, delirium tremens."

This was an immense relief—much better DTs than a schizo-phrenic psychosis. But I was quite aware of the dangers of the DTs: confusion, disorientation, hallucination, delusion, dehydration, fever, rapid heartbeat, exhaustion, seizures, death. I would have advised anyone else in my state to get to an emergency room im-mediately, but for myself I wanted to tough it out and experience it to the full. Carol agreed to sit with me for the first day and then, if she thought I was safe by myself, she would look by or phone me at intervals, calling in outside help if she judged it necessary. Given this safety net, I lost much of my anxiety and could even, in a way, enjoy the phantasms of delirium tremens (though the myriads of small animals and insects were anything but pleasant). The hal-lucinations continued for almost ninety-six hours, and when they finally stopped I fell into an exhausted stupor.

As a boy, I had known extreme delight in the study of chemistry and the setting up of my own chemistry lab. This delight seemed to desert me at the age of fifteen or so; in my years at school, university, medical school, and then internship and residency, I kept my head above water, but the subjects I studied never excited me in the same intense way as chemistry had when I was a boy. It was not until I arrived in New York and began seeing patients in a migraine clinic in the summer of 1966 that I began to feel a little stirring of the intellectual excitement and emotional engagement I had known in my earlier years. In the hope of whipping up these intellectual and emotional excitements even further, I turned to amphetamines.

I would take the stuff on Friday evenings after getting back from work and would then spend the whole weekend so high that images and thoughts would become rather like controllable hal-lucinations, imbued with ecstatic emotion. I often devoted these "drug holidays" to romantic daydreaming, but one Friday, in Feb-ruary 1967, while I was exploring the rare-book section of the medical library, I found and took out a rather rare book on mi-graine entitled *On Megrim, Sick-Headache, and Some Allied Disorders: A Contribution to the Pathology of Nerve-Storms*, written in 1873 by one Edward Liveing, MD. I had been working for several months in a migraine clinic, and I was fascinated by the range of symptoms and phenomena that could occur in migraine attacks. These attacks often included an aura, a prodrome in which aberrations of perception and even hallucinations occurred. They were entirely benign and would last only a few minutes, but those few minutes provided a window onto the functioning of the brain and how it could break down and then reintegrate. In this way, I felt, every attack of migraine opened out into an encyclopedia of neurology.

I had read dozens of articles about migraine and its possible basis, but none of them seemed to present the full richness of its phenomenology or the range and depth of suffering that patients might experience. It was in the hope of finding a fuller, deeper, and more human approach to migraine that I took out Liveing's book from the library that weekend. So, after downing my bitter draft of amphetamine—heavily sugared, to make it more palatable—I started reading. As the intensity of the amphetamine effect took hold of me, stimulating my emotions and imagination, Liveing's book seemed to increase in intensity and depth and beauty. I wanted nothing but to enter Liveing's mind and imbibe the atmosphere of the time in which he worked. In a sort of catatonic concentration so intense that in ten hours I scarcely moved a muscle or wet my lips, I read steadily through the five hundred pages of *Megrim*. As I did so, it seemed to me almost as if I were becoming Liveing himself, actually seeing the patients he described. At times I was unsure whether I was reading the book or writing it. I felt myself in the Dickensian London of the 1860s and '70s. I loved Liveing's humanity and social sensitivity, his strong assertion that migraine was not some indulgence of the idle rich but could affect those who were poorly nourished and worked long hours in ill-ventilated factories. In this way, his book reminded me of Henry Mayhew's great 1861 study of London's working classes, but equally one could tell how well Liveing had been trained in biology and the physical sciences, and what a master of clinical observation he was. I found myself thinking, This represents the best of mid-Victorian science and medicine; it is a veritable masterpiece! The book gave me what I had been hungering for during the months that I was seeing patients with migraine and being frustrated by the thin, impoverished articles that seemed to constitute the modern "literature" on the subject. At the height of this ecstasy, I saw migraine shining like an archipelago of stars in the neurological heavens.

But about a century had passed since Liveing worked and wrote in London. Rousing myself from my reverie of being Liveing or one of his contemporaries, I came to and said to myself, "Now it is the 1960s, not the 1860s. Who could be the Liveing of our time?" A disingenuous clutter of names spoke themselves in my mind. I thought of Dr. A. and Dr. B. and Dr. C. and Dr. D., all of them good men but none with that mixture of science and humanism which was so powerful in Liveing. And then a very loud internal voice said, "You silly bugger! You're the man!"

On every previous occasion when I had come down after two days of amphetamine-induced mania, I had experienced a severe reaction in the other direction, feeling an almost narcoleptic drowsiness and depression. I would also have an acute sense of folly that I had endangered my life for nothing—amphetamines in the large doses I took would give me a sustained pulse rate close to 200 and a blood pressure of I-know-not-what; several people I knew had died from overdoses of amphetamines. I would feel that I had made a crazy ascent into the stratosphere but had come back empty-handed and had nothing to show for it; that the experience had been as empty and vacuous as it was intense. This time, though, when I came down, I retained a sense of illumination and insight; I had had a sort of revelation about migraine. I had a sense of resolution, too, that I was indeed equipped to write a Liveing-like book, that perhaps I could be the Liveing of our time.

The next day, before I returned Liveing's book to the library, I photocopied the whole thing, and then, bit by bit, I started to write my own book. The joy I got from doing this was real—infinitely more substantial than the vapid mania of amphetamines—and I never took amphetamines again.

ELIZABETH KOLBERT

Recall of the Wild

FROM *The New Yorker*

FLEVOLAND, WHICH SITS more or less in the center of the Netherlands, half an hour from Amsterdam, is the country's newest province, a status that is partly administrative and partly existential. For most of the past several millennia, Flevoland lay at the bottom of an inlet of the North Sea. In the 1930s a massive network of dams transformed the inlet into a freshwater lake, and in the 1950s a drainage project, which was very nearly as massive, allowed Flevoland to emerge out of the muck of the former sea floor. The province's coat of arms, drawn up when it was incorporated in the 1980s, features a beast that has the head of a lion and the tail of a mermaid.

Flevoland has some of Europe's richest farmland; its long, narrow fields are planted with potatoes and sugar beets and barley. On each side of the province is a city that has been built from scratch: Almere in the west and Lelystad in the east. In between lies a wilderness that was also constructed, Genesis-like, from the mud.

Known as the Oostvaardersplassen, a name that is pretty much unpronounceable for English-speakers, the reserve occupies 15,000 almost perfectly flat acres on the shore of the inlet-turned-lake. This area was originally designated for industry; however, while it was still in the process of drying out, a handful of biologists convinced the Dutch government that they had a better idea. The newest land in Europe could be used to create a Paleolithic landscape. The biologists set about stocking the Oostvaardersplassen with the sorts of animals that would have inhabited the region in prehistoric times—had it not at that point been underwater. In many cases, the animals had been exterminated, so they had to settle for the next best thing. For example, in place of the aurochs, a large and now extinct bovine, they brought in Heck cattle, a variety specially bred by Nazi scientists. (More on the Nazis later.) The cattle grazed and multiplied. So did the red deer, which were trucked in from Scotland, and the horses, which were imported from Poland, and the foxes and the geese and the egrets. In fact, the large mammals reproduced so prolifically that they formed what could, with a certain amount of squinting, be said to resemble the great migratory herds of Africa; the German magazine *Der Spiegel* has called the Oostvaardersplassen "the Serengeti behind the dikes." Visitors now pay up to forty-five dollars each to take safari-like tours of the park. These are especially popular in the fall, during rutting season.

Such is the success of the Dutch experiment—whatever, exactly, it is—that it has inspired a new movement. Dubbed Rewilding Europe, the movement takes the old notion of wilderness and turns it inside out. Perhaps it's true that genuine wildernesses can only be destroyed, but new "wilderness," what the Dutch call "new nature," can be created. Every year, tens of thousands of acres of economically marginal farmland in Europe are taken out of production. Why not use this land to produce "new nature" to replace what's been lost? The same basic idea could, of course, be applied outside of Europe—it's been proposed, for example, that depopulated expanses of the American Midwest are also candidates for rewilding.

I visited the Oostvaardersplassen during a stretch of very blue days in early fall. As it happened, two film crews, one Dutch and the other French, were also there. The French crew, whose credits include the international hit *Winged Migration*, was scouting the reserve for possible use in an upcoming feature about the history of Europe as seen through the eyes of other species. The Dutch crew was finishing up a full-length nature documentary. One afternoon we all got into vans and drove to the middle of the park. A stiff breeze was blowing, as it almost always does near the North Sea. We passed a marshy area covered in reeds, which nodded in the wind. Ducks bobbed in a pond. Farther on, where the land grew drier, the reeds gave way to grass. We passed a herd of red deer, some aurochs wannabes, and the carcass of a deer, which had been picked almost clean by foxes and ravens. (The Dutch crew had filmed the scavenging with a time-lapse camera.) Eventually, we came to a herd of about a thousand wild—or at least feral—horses. They whinnied and cantered and shook their heads. The horses were an almost uniform buff color, and the breeze lifted their manes, which were dark brown. We all piled out of the vans. The horses seemed not to notice us, though we were just a few yards away.

"Ah, c'est joli ça!" the French exclaimed. A flock of black-and-white barnacle geese rose into the air and then, a moment later, a yellow train clicked by, carrying passengers from Almere to Lelystad or perhaps vice versa. A few members of the French crew had brought along video cameras. As they panned across the horses—at the edge of the herd, a mare nuzzled a foal that couldn't have been more than two or three days old—I wondered what they would do with the high-voltage power lines in the background. It occurred to me that, like so many postmodern projects, the Oostvaardersplassen was faintly ridiculous. It was also, I had to admit, inspiring.

If one person could be said to be responsible for the Oostvaardersplassen, it is an ecologist named Frans Vera. Vera, who is sixty-three, has gray hair, a gray beard, and a cheerfully combative manner. He spent most of his adult life working for one or another branch of the Dutch government and now works for a private foundation, of which, as far as I could tell, he is the sole employee. Vera picked me up one day at my hotel in Lelystad, and we drove over to the reserve's administrative offices, where we had a cup of coffee in a room decorated with the mounted head of a very large, black Heck bull.

Vera explained that he first became interested in the Oostvaardersplassen in the late 1970s. At that point, he had just graduated from university in Amsterdam and was unemployed. He read an article about some graylag geese that had appeared in the reclaimed area, which was then a boggy no man's land. The geese kept the vegetation low by chomping on it, and in this way maintained their marshy habitat. Vera was an avid bird watcher, and the story intrigued him. He wrote his own article, arguing that the place ought to be turned into a nature preserve. Soon afterward, he got a job with the Dutch forestry agency.

In the late seventies, the prevailing view in the Netherlands was —and, to a certain extent, it still is—that nature was something to be managed, like a farm. According to this view, a preserve needed to be planted, pruned, and mowed, and the bigger the preserve, the more intervention was required. Vera chafed at this notion. The problem, he decided, was that Europe's large grazers had been hunted to oblivion. If they could be restored, then nature could take care of itself. This theory, coming from a very junior civil servant, was not particularly popular.

"Mostly there's no trouble as long as you are within the borders of an accepted paradigm," Vera told me. "But be aware when you start to discuss the paradigm. Then it starts to be only twenty-five percent discussion of facts and seventy-five percent psychology. The thing I most often heard was, 'Who do you think you are?'" Undaunted, Vera kept pushing. He had a few allies at various government ministries, and one of them arranged for him to get the money to buy some Heck cattle. In 1983, while the future of the Oostvaardersplassen was still being debated, Vera acquired the cows from Germany, although he had not yet secured permission from the governing authorities to release them.

"I bought them and I was standing here with the trucks," he recalled happily. "And they were so angry!" This first group of Heck cattle was not allowed onto the site, but a second group, acquired some months later, was let in. The following year, Vera bought forty Konik horses from Poland. Koniks are believed to be descended from tarpans, one of the world's last subspecies of truly wild horse, which survived in eastern Europe into the nineteenth century. (Practically all the horses that are called "wild" today are, in fact, the offspring of domesticated horses that were, at some point or another, let loose.) Red deer, which are closely related to what Americans call elk, were brought in during the 1990s.

Meanwhile, other animals were finding their way to the Oost-vaardersplassen on their own. Foxes arrived, as did muskrats, which in Europe count as an invasive species. Buzzards and gos-hawks and gray herons and kingfishers and kestrels turned up. A pair of very large white-tailed eagles swooped in and built their nest in an improbably small tree. In 2005 a rare black vulture appeared, but after a few months in residence it wandered onto the railroad tracks, where it was hit by a train. (The rail line runs along the southern edge of the preserve.) Vera's dream is that one

day the Oostvaardersplassen will be connected to other nature reserves in the Netherlands—a plan that has been partly but never
fully funded—and that this will, in turn, allow it to attract wolves.
Wolves were extirpated from most of western Europe more than a
century ago, but owing to stringent protections put in place over
the past few decades, they have recently been making a comeback
in countries like Germany and France. (Two packs, with about ten
wolves each, now live within forty miles of Berlin.) Last year a wolf
believed to be the first seen in Holland since the 1860s was spotted
about seventy miles southeast of the Oostvaardersplassen, in the
town of Duiven.

"That is probably unimaginable for people in the United States
—having wolves in the Netherlands," Vera said. "But it is the future."

After we had finished our coffee, we got into a truck and drove
through the gates of the preserve. So effectively have the cows
and the horses and the deer kept the place grazed that there was
barely a bush to be seen—just acre after very flat acre of clipped
grass, like a bowling green. We passed a few groups of deer and a
fox that looked back at us with pale, glittering eyes. Vera stopped
the truck at a lookout built on stilts. We climbed up a narrow ladder. "This is a window that shows us how the Netherlands looked
thousands of years ago," he said, gesturing at the grassland below.

A corollary of Vera's theory about large grazers is a second hypothesis, which he has pushed even more vigorously than the first,
if that's possible. Among ecologists, the prevailing view of Europe
in its natural, which is to say preagrarian, state is that it was heavily forested. (The continent's last stands of old-growth forest are
found on the border of Poland and Belarus, in the Białowieża Forest, which the author Alan Weisman has described as a "relic of
what once stretched east to Siberia and west to Ireland.") Vera
argues that even before Europeans figured out how to farm, the
continent was more of a parklike landscape, with large expanses
of open meadow. It was kept this way, he maintains, by large herds
of herbivores—aurochs, red deer, tarpans, and European bison.
(The bison, also known as wisents, were hunted nearly to extinction by the late 1800s.)

Vera has written up his argument in a dense, 500-page treatise
that has received a good deal of attention from European natural-

ists, not all of it favorable. A botany professor at Dublin's Trinity College, Fraser Mitchell, has written that an analysis of ancient pollen "forces the rejection of Vera's hypothesis." Vera, for his part, rejects the rejection, arguing that precisely because they ate so much grass, the aurochs and the wisents skewed the pollen record. "That is a scientific debate that is still going on," he told me.

Like the rest of Flevoland, the Oostvaardersplassen lies about fifteen feet below sea level and is protected from flooding by a series of thick earthen dikes. As a result, when you are standing in the park, the lake, known as the Markermeer, is above you, which produces the vertiginous sense of a world upside down. In the lovely weather, the Markermeer was filled with sailboats; these seemed to be hovering above the horizon, like zeppelins.

"What we see here is that instead of what many nature conservationists think—that something that is lost is lost forever—you can have the conditions to have it redeveloped," Vera told me. "So this is the ultimate proof. There's no bird here who says, 'I won't breed here, because it's unnatural—it's four and a half meters below sea level, and I never did that.'" We drove on and stopped to take a look at the nest built by the white-tailed eagles, another animal that only very narrowly avoided extinction. The eagles showed up in the Oostvaardersplassen in 2006 and became the first pair to breed in the Netherlands since the Middle Ages. Their nest —empty at the time of my visit—was an extraordinary structure, made out of sticks and nearly the size of an armchair. It seemed ready to topple the scrawny tree it was perched in. Vera was particularly pleased with the eagles, because several ornithologists had told him the birds would nest only in very tall, mature trees, of which the Oostvaardersplassen has none.

"Many so-called specialists thought this would be impossible," he said. "The eagles had a different opinion."

Access to the Oostvaardersplassen by humans is strictly controlled, and that morning neither of the film crews was there and no tours were out, so Vera and the animals and I pretty much had the place to ourselves. The quiet was interrupted only by the squawking of the geese and the clatter of an occasional train. We continued west, skirting a herd of red deer. A dead horse was lying in the middle of the herd. Its chest was bloated, and there was a large dark hole where its anus once had been. Vera speculated that it had been made by foxes trying to get at the horse's entrails.

Like genuinely wild animals, those in the Oostvaardersplassen are expected to fend for themselves. They are not fed or bred or vaccinated. Also like wild animals, they often die for lack of resources; for the large herbivores in the reserve, the mortality rate can approach 40 percent a year. From a public relations point of view, this is far and away the most controversial aspect of Vera's scheme. When the weather is harsh, there's widespread starvation in the preserve, which provides gruesome images for Dutch TV. Often the dying animals are shown huddled up against the fences of the Oostvaardersplassen, a scene that invariably leads to comparisons with the Holocaust.

"You can't have a discussion without the Second World War coming up," Vera told me. "It's really sick-making." In the fall of 2005, the controversy became so heated that the Dutch government appointed a committee—the International Committee on the Management of Large Herbivores in the Oostvaardersplassen, or ICMO—to look into the matter. ICMO recommended a policy of "reactive culling," under which the animals would be monitored over the winter, and those that seemed too weak to survive until spring would be shot.

Michael Coughenour, a research scientist at the Natural Resource Ecology Laboratory at Colorado State University, was a member of ICMO. He told me that while it was difficult to compare mortality rates at the Oostvaardersplassen to those in a place like the Serengeti, "severe-winter die-offs are a natural thing."

"I didn't see anything that looked bad to me," he went on, referring to a visit the committee members made to the Oostvaardersplassen. "I think it's a great experiment to let it run and see what happens."

Even though ICMO's recommendations were adopted, many critics were not satisfied, and in 2006 a Dutch animal-welfare association sued the managers of the Oostvaardersplassen for what it alleged was continuing mistreatment. The group lost the case, appealed, and lost again. Then, in the winter of 2010, an unusually cold one in northern Europe, a Dutch news program aired a segment on the Oostvaardersplassen that showed an emaciated deer stumbling into a half-frozen pond and drowning. A public outcry ensued, prompting an "emergency" debate in parliament.

"It's an illusion to think we can go back to primordial times, dressed in bear furs and floating around in hollowed-out trees," the MP who led the debate, Henk Jan Ormel, said. "The world of today looks very different, and we shouldn't make the animals of the Oostvaardersplassen bear the burden of this."

"It became political," Sip van Wieren, a professor of ecology at Wageningen University, told me. "*Very* political." A second ICMO was convened. This one recommended a policy of "early reactive culling," under which the animals that were deemed unlikely to survive the winter would be shot in the fall. How exactly the rangers at the Oostvaardersplassen were supposed to figure out in November which animals would be starving by February was left rather vague.

When I visited, in September, the number of grazers in the park was at its annual peak, with more than 3,000 deer, 1,000 horses, and 300 Heck cattle. Eventually, it is hoped, birthrates in the Oostvaardersplassen will decline, and the population will reach some kind of equilibrium, but in the meantime the shooting continues. Vera and I came upon a group of cows sunning themselves near a dead tree. They regarded us warily, through glassy black eyes. The adults looked fearfully robust, but some of the calves seemed a bit shaky; within a few months, I figured, they'd probably be carcasses. Vera told me that he viewed "early reactive culling" as an arrangement whose only real beneficiaries were humans; as far as the ungulates were concerned, he thought, starving to death was a very peaceful way to go.

"It only has to do with the acceptance of people," he said, "and nothing, in my mind, to do with the suffering of animals."

There are more than 1.5 billion cows in the world today, and all of them are believed to be descended from the aurochs—*Bos primigenius*—which once ranged across Europe, much of Asia, and parts of the Middle East. Aurochs were considerably more impressive beasts than domesticated cattle. Julius Caesar described them as being just "a little below the elephant in size," with "strength and speed" that was "extraordinary." (It is unlikely that he ever actually saw one.) More recent estimates suggest that males were nearly six feet high at the withers and females five feet. By Roman times, humans had so diminished the aurochs' numbers that the animals were missing from most of their former habitat.

By the 1500s, the only place they could still be found in the wild was in the Polish Royal Forests, west of Warsaw. The animals there were understood to be extremely rare, and special game-keepers were hired to protect them. But their numbers continued to dwindle. In 1557 some fifty aurochs were counted. Forty years later, only half that many remained, and by 1620 only one aurochs —a female—was left. She died in 1627. The aurochs thus earned, as the Dutch writer Cis Van Vuure has put it, "the dubious honor of being the first documented case of extinction." (The next case was the dodo, four decades later.)

The aurochs was essentially forgotten until the early twentieth century, when a spate of scientific papers on the animal appeared. In the 1920s, two German brothers, Heinz and Lutz Heck, both zoo directors, decided to try to breed back the aurochs, using the genetic material that had been preserved in domesticated cattle. This was, of course, long before DNA testing—or even the discovery of DNA. To guide their efforts, the brothers mainly relied on old pictures of aurochs, many of them drawn by people with no firsthand knowledge of the animal. The brothers chose different kinds of cows for their breeding efforts: Heinz, who directed the zoo in Munich, crossed, among other breeds, Scottish Highland cattle and German Anglers, while Lutz, the director of the Berlin zoo, mixed Spanish fighting cattle with Corsican and Camargue cattle. Nevertheless, the two claimed that their efforts had produced similar results, which, they argued, proved that "the fundamental principle of breeding back was correct." Even though he continued to crossbreed his crossbreeds, Heinz decided that the project had been successfully completed. "The wild bull, the aurochs, lives again," he wrote.

Not long afterward, the project became tangled up in German politics. In 1938 Lutz, a committed Nazi, was appointed to the Third Reich's Forest Authority. His idea of breeding back the aurochs dovetailed neatly with the Nazis' scheme of restoring Europe, through selective human breeding, to its mythic Aryan past. Lutz sent some of his "aurochs" to the Rominten Heath in East Prussia—now Poland—where Hermann Göring had his favorite hunting lodge. Other Heck-bred cows were installed on the grounds of Göring's estate north of Berlin. Most—perhaps all—of these animals were killed toward the end of World War II. (According to Clemens Driessen, a Dutch academic who has studied the Heck brothers, Göring personally shot some of the cattle on his estate as the Soviets bore down on Berlin.) But some Heck cattle at the Munich zoo and in parks in Augsburg, Münster, and Duisburg survived.

Over the years, even as Heck cattle have been raised uneventfully in once Nazi-occupied nations like the Netherlands—it's the descendants of the Munich-bred cows that now graze the Oostvaardersplassen—they've never managed to shake their Fascist associations. Many regard them as a sort of veterinary version of the Hitler Diaries—half horror, half joke. Not long ago, when a British farmer imported some Heck cattle from Belgium, the story made national news.

NAZI "SUPER-COWS" SHIPPED TO DEVON FARM, the *Guardian* reported.

THE HERD REICH, ran the headline in the *Sun*.

As more aurochs remains have been unearthed and more sophisticated research has been done on them, it's become clear that the Heck brothers' creation is a far cry from the original— Heck cattle are too small, their horns have the wrong shape, and the proportions of their bodies are off. All of which has led to a new, de-Nazified effort to back-breed the aurochs. This project is based in the Dutch city of Nijmegen, about fifty miles southeast of Amsterdam, and is entirely independent of the Oostvaardersplassen. Still, it reflects much the same can-do, "what is lost is not lost forever" approach to conservation. So while I was in the Netherlands I decided to go for a visit.

"Watch out," Henri Kerkdijk warned. It was another surprisingly blue day, and we were tromping through a weedy field toward a line of trees. I looked back at him, which turned out to be a mistake, because at that moment I stepped into a large pile of cow shit. As I scraped it from my shoes, I wondered how much bigger the pile would have been had it been produced by an actual aurochs.

Standing in the shade of the trees were about a dozen cows of varying color and size. Kerkdijk pointed to two black bulls bent over a patch of grass. The first was called Manolo Uno. He was two years old and not yet fully grown, but already he measured almost five feet at the withers. He had a grayish muzzle, a light stripe down his back, and forward-tilting horns that reminded me of Ferdinand's. I have no idea how closely he resembled an actual aurochs; certainly, though, he seemed a very imposing beast,

larger and more menacing-looking than the Heck cattle at the
Oostvaardersplassen. The second bull, Rocky, was a year younger
than Manolo but almost as big. This Kerkdijk took as a particularly
promising sign. "That one's going to be really tall," he said.

Four years ago, Kerkdijk teamed up with an environmental consultant named Ronald Goderie to start the TaurOs program, the
stated goal of which is to give "the rebuilding of the aurochs a serious try." (In a recent write-up of the effort, the two men dismiss
Heck cattle as "considered by experts to be a failure.") At the point
that I met with them, the project had generated nearly a hundred
calves, of which Manolo Uno and Rocky had been deemed the
most aurochs-like. To create the calves, Kerkdijk and Goderie had
crossed several so-called primitive cattle breeds—varieties developed hundreds, even thousands, of years ago, and therefore more
likely to retain aurochs-like features. Manolo, for example, represents a cross between an Italian breed known as Maremmana
primitivo and a Spanish breed known as Pajuna. At two, he was old
enough to be crossbred himself. But he had refused to part with
any of his semen for the purpose of artificial insemination, a de-
murral that Kerkdijk took as evidence of his virility and a further
positive sign.

Ninety years after the Heck brothers' attempt, the basic idea
behind back-breeding remains pretty much the same. If different
breeds of primitive cattle preserve different stretches of the aurochs's genetic material, then reassembling those stretches should
produce something close to—though not exactly like—the original. (Kerkdijk and Goderie have decided that their new animal
should be called not an aurochs but a "tauros.") Scientists in England and Ireland have succeeded in sequencing a small subset of
the aurochs's DNA—its mitochondrial DNA—using a 7,000-year-
old bone that was found in a cave in Derbyshire. Other scientists
have been approached about sequencing the entire genome.
When—or, really, if—this work is completed, it should be possible
to gauge how close a calf comes to an authentic aurochs by analyzing a blood sample or a bit of saliva.

According to the timetable Kerkdijk and Goderie have drawn
up, herds of "tauroses" should be ready by around 2025. By that
point, the two expect that large tracts of Europe will have been
rewilded, and the animals will be allowed to roam across them.
How the intervening years' worth of breeding and crossbreeding

and genetic evaluation will be funded remains a bit murky. Currently, the project is supported in part by renting cows to nature parks and in part by butchering them. The meat is marketed as "wild beef," and it commands a premium in Amsterdam, where it is available only to customers who sign up for delivery in advance. Kerkdijk said that "wild beef" sales had risen dramatically over the last year or so, owing to interest in the tauros. I asked if I could try some.

"Did you bring your bow and arrow?" Goderie asked.

Like so much in Europe today, the term "rewilding" is an American import. It was coined in the 1990s and first proposed as a conservation strategy by two biologists, Michael Soulé, now a professor emeritus at the University of California at Santa Cruz, and Reed Noss, a research professor at the University of Central Florida. According to Soulé and Noss, the problem with most conservation plans was that they aimed to protect what exists. Yet what exists is often just a shadow of what once was. In most of the United States, large predators like wolves and cougars have been wiped out. Without top predators, the two argued, ecosystems no longer really function as systems.

"A cynic might describe rewilding as an atavistic obsession," they wrote. "A more sympathetic critic might label it romantic. We contend, however, that rewilding is simply scientific realism." According to Soulé and Noss, rewilding demanded, in addition to predators, the establishment of large, strictly protected "core" reserves and migratory corridors linking one to the next. They summarized their formula as "the three C's: cores, corridors, and carnivores." These ideas are now considered mainstream by conservation biologists, even those who would not necessarily describe themselves as proponents of rewilding.

In 2005 a dozen biologists took the concept of rewilding one step further. In an article published in the journal *Nature,* the group presented a plan for what it called "Pleistocene rewilding."

When humans arrived in North America some 13,000 years ago, toward the end of the last ice age, they killed off most of the continent's largest mammals, leaving key ecological roles unfilled. The Pleistocene rewilders proposed finding substitute animals that could serve in their place. For instance, African or Asian elephants could be let loose to make up for the long-lost woolly mammoth.

Similarly, Bactrian camels, which are native to the steppes of Central Asia, could take up the slack left by the vanished North American *Camelops*. The authors—almost all of them were academics—envisioned a series of small-scale experiments leading up to the creation of "one or more 'ecological history parks,'" which would cover "vast areas of economically depressed parts of the Great Plains." In these huge "history parks," elephants, camels, and African cheetahs—to replace the missing American cheetah—would roam freely. The ecologists called their plan "an optimistic alternative" to what was otherwise likely to be a future filled with "ever more pest-and-weed-dominated landscapes" and "the extinction of most, if not all, large vertebrates."

The lead author of the *Nature* article, Josh Donlan, now runs a nonprofit group called Advanced Conservation Strategies and is a visiting fellow at Cornell. He characterized reactions to Pleistocene rewilding as "bimodal."

"People either loved it or hated it, both in the scientific community and in the public," he told me. In the United States, Pleistocene rewilding never got very far; the only practical step that's been taken has been the reintroduction to private land in New Mexico of a giant tortoise known as the Bolton tortoise. (The Bolton tortoise, which disappeared from what's now the United States about 8,000 years ago, survived south of the border in very small numbers.) As it happened, though, a Russian scientist named Sergey Zimov had a similar idea. Also in 2005, he published an article in *Science* describing an experimental preserve in Siberia that he had set up and named the Pleistocene Park. Zimov's aim was to show that the area, which 10,000 years or so ago supported great herds of large mammals, was still capable of doing so.

"We are not trying exactly to reconstruct the mammoth steppe ecosystem, because we don't have the mammoth," Zimov told me recently by phone from St. Petersburg. "But we are trying to reconstruct the highly productive steppe ecosystem." Zimov brought in reindeer and a breed of very cold-hardy horses known as Yakutians. A few years ago, he imported five European bison to the park, but only one—a male—survived the second winter. "Now we are looking for girlfriends," Zimov said. Several musk oxen were also brought in, but they, too, were all males. "We also search females for them," Zimov told me. The Pleistocene Park, which is in northeastern Siberia, is so remote that almost no one who isn't conducting research there has ever visited it.

As Europeans have taken up the term, "rewilding" has shifted its meaning yet again. The concept has become at once less threatening and more gastronomically appealing: it is expected that visitors to the continent's rewilded regions will be able to enjoy not just the safari-like tours but also the local cuisine. (One park in Portugal in the process of "rewilding" offers its own brand of olive oil.)

Rewilding Europe, the group that is pushing the concept most vigorously, was founded three years ago by two Dutchmen, a Swede, and a Scot. One of the Dutchmen, Wouter Helmer, lives not far from the field where Manolo and Rocky are pastured, and the day after I visited the bulls I went to meet him at his house, which is at the edge of a park, in a small clearing that made me think of Goldilocks.

Helmer explained that the goal of Rewilding Europe was, in effect, to create giant versions of the Oostvaardersplassen, each at least fifteen times as large. "Frans Vera always says, 'If the Dutch can do it, everyone can do it,'" he told me. To get the project started, the group has raised more than 6 million euros—roughly $7.5 million dollars—much of it from the Dutch postcode lottery, which might be compared to the New York State lottery, except that the proceeds go to charity. Last year, after receiving twenty applications from organizations across the continent, the group chose five regions to serve as what it calls "model rewilding areas" —a part of the Danube delta spanning the border of Romania and Ukraine; an area in the southern Carpathian Mountains, also known as the Transylvanian Alps; and areas in the eastern Carpathians, the mountains of Croatia, and the western Iberian Peninsula. One quality these areas share is that fewer and fewer people want to live in them.

"There's no economy in big parts of Europe," Helmer told me. "We think it's a window of opportunity." The idea is to rewild the areas by connecting existing reserves with tracts of abandoned land and working farms whose owners can be persuaded to let a herd of aurochs (or tauroses) wander across their property. (The lure for landowners is supposed to be an influx of tourists, who will come and open their wallets.)

Helmer stressed to me that Rewilding Europe was not particularly concerned about whether the new landscape that would be created would resemble the ancient one that had been altered or destroyed. "We're not looking backward but forward," he said at one point.

"We try to avoid too much discussion of wilderness," he observed at another. "For us, that is not the most important thing —at the end will this be a wilderness or not? It will be wilder than it was, and that's what matters."

One morning not long after this, I found myself sitting in a small hut, staring at a pile of dead chickens. The chickens had pure white feathers that were matted with blood, and they lay with their half-severed heads and rigid legs tilted at grotesque angles. After a while, a half dozen Griffon vultures settled into a nearby tree. Griffon vultures are large birds with light-colored faces and dark bodies, and the group in the tree resembled a gathering of harpies. A little while later, a pair of black vultures showed up and began circling overhead. Black vultures are even larger than Griffons, with wingspans that can reach ten feet. They are majestic, funereal-looking birds, and watching them feels like a premonition of one's own death. The chickens had been laid out as part of a supplementary feeding program for the birds, who, it seemed, were not hungry. The black vultures continued to circle, the Griffon vultures continued to sit in the tree, and the small hut grew stuffier. After a few hours, my companion, Diego Benito, decided that the spectacle we had come to see was not going to take place, and so, disappointed, we left.

Benito runs a 1,300-acre nature preserve in far western Spain called the Campanarios de Azaba. The preserve is part of the Rewilding Europe "model area" in western Iberia, and of the five areas it's the easiest to get to. Nevertheless, the trip there involves a four-hour drive from Madrid, through the provinces of Ávila and Salamanca.

Since the vultures weren't cooperating, Benito suggested we tour the rest of the reserve. Until fairly recently, the place had been a farm, and it was dotted with oak trees whose acorns had gone to fattening pigs. It was hot and dry as we crunched along through the underbrush. Even though I knew the nearest town wasn't more than a few miles away, the terrain seemed empty enough to get lost in, and I was reminded of a time in the New Mexico desert when I'd read a trail map wrong and found myself walking in circles. We encountered some very handsome horses, which, Benito told me, belonged to a rare and ancient Spanish breed known as Retuertas. Farther on we came to a fenced-in area filled with a network of small but clearly man-made tunnels. These, Benito explained, had been dug for the benefit of rabbits, which in Spain—and, indeed, throughout Europe—have been decimated by a disease known as myxomatosis. The myoma virus was purposefully introduced on a private estate in France as a rabbit-control measure in the 1950s and has since spread across the continent. (The loss of rabbits has led to a decline in animals that prey on rabbits, like the Iberian lynx, which is now considered to be critically endangered.) The fences were supposed to protect some reintroduced rabbits from foxes, but the rabbits had refused to stay put, so now the enclosures were empty. The same was true of a series of circular platforms that had been erected in some oak trees as nesting sites for black storks. The black storks hadn't been interested in them.

"You can't be a hundred percent sure of success, because wild animals are wild animals," Benito told me. We went looking for some Sayaguesa cows that had recently been purchased with Rewilding Europe money, but they seemed to be avoiding us. Sayaguesas are another primitive breed of interest to the TaurOs program, an enterprise that Benito told me he was eager to get involved in. "If you want to sell a product, you have to have a story," he said.

That afternoon, after a lunch of local (and quite tasty) pork cutlets, we drove out of the reserve to the top of a nearby mountain. Along the way, we passed through a couple of villages that, Benito explained, were in the process of disappearing; the schools had closed for lack of children and only the old people remained. In one of the towns, La Encina, we stopped to meet the mayor, a slight, elderly man named José Maria. According to Maria, the number of residents in La Encina had dropped by more than 50 percent in just the past fifteen years. He was enthusiastic about the idea of rewilding, he said, because it had "a lot of potential to bring tourists." From the top of the mountain, we could see across to Portugal, some fifteen miles away. The valley was a patchwork of brown fields, pine forests that had been planted during the Franco era, and evenly spaced oaks of the sort I'd seen at the preserve. According to a brochure that Wouter Helmer had given me, the entire region was ripe for rewilding, owing to "rural depopulation"; the aim was to transform at least 1,000 square kilometers, or 250,000 acres. I tried to imagine the whole valley converted into an Iberian version of the Oostvaardersplassen. Certainly it was a lot less populated than the outskirts of Amsterdam. Still, I realized, I wasn't sure what I was supposed to be envisioning. The pine plantations could never be considered wild: would they have to go? What about the pruned oaks, and the pigs that were still snuffling around them for acorns, and the brown fields, and all the tiny, dying towns waiting for an influx of tourists?

One of the appeals of rewilding is that it represents a proactive agenda—as Josh Donlan and his Pleistocene rewilding colleagues put it, a hopeful alternative to just sitting around, mourning what's been lost. In a rewilded world, even extinction need not be considered irrevocable; the aurochs will lie down with the lynx, and the deer and the elephants will roam. On a planet increasingly dominated by people—even the deep oceans today are being altered by humans—it probably makes sense to think about wilderness, too, as a human creation. The more I saw, the more I understood why Europeans, in particular, were attracted to the idea, and the more I wanted to be convinced that it could work. But, as I looked back toward the Campanarios de Azaba, I thought of the vacant rabbit tunnels and the empty platforms built for the storks, and I wasn't at all sure.

It was dusk by the time we headed down the mountain. Benito got a call on his cell phone from a local farmer who had a dead pig he thought the vultures might be interested in. On our way back, we stopped by to see what had happened to the chickens. Every one of them was gone, including the bones.

KEITH GESSEN

Polar Express
FROM *The New Yorker*

THE ICE-CLASS BULK carrier *Nordic Odyssey* docked at the port of Murmansk, Russia, just after six in the morning on July 5, 2012. It had a green deck and a red hull, and was 738 feet long, 105 feet wide, and 120 feet from top to bottom; empty, it weighed 14,000 tons. It was an eighty-story building turned on its side and made to float. The *Odyssey* had come to pick up 65,000 tons of iron ore and take it to China via the Northern Sea Route—through the ice of the Arctic seas and then down through the Bering Strait.

Murmansk, which rises along one bank of a fjord thirty miles south of the Barents Sea, is the world's largest city north of the Arctic Circle, and yet as soon as a visitor got past harbor security at the gate, the city disappeared. The pier was covered by huge mounds of coal and iron ore. Train cars kept pulling in with more; tall yellow cranes dipped into them and deposited the ore onto the mounds, and then the train cars pulled out again. It was as if Russia were coughing up her insides. The cranes' grabs could barely squeeze into the railcars. The deep, rumbling sounds of steel on steel echoed in the quiet of the fjord.

The *Odyssey* is owned by a Danish shipping company called Nordic Bulk. In 2010 the company was asked to get a load of ore from Norway to China. The normal route would be either south through the Suez Canal or even farther south, around the Cape of Good Hope, but the Suez route would take you by the coast of Somalia, home to the world's most enterprising pirates, and the Cape Hope route would take too long. Mads Petersen, the co-chairman of Nordic Bulk, wondered if there was another way. As it happened, the shortest route from Norway to China was through the Arctic. "And I thought, Maybe the Northern Sea Route has opened up, because of global warming," Petersen said, recounting his thought process two years later in Murmansk. He is just past thirty, gregarious, and big—six-foot-two and 260 pounds. I said, "You started going to the Arctic because you read an article about global warming?"

Petersen shook his head. "In Denmark, you do not 'read an article' about global warming," he said. "You hear about it, all the time."

Petersen contacted Rosatomflot, the state company that owns Russia's six nuclear icebreakers (the largest such fleet in the world), and made a deal to send his cargo through the Arctic with an icebreaker escort. The price was $300,000, but the projected savings in fuel and time would make up for it and then some. Moreover, it was an adventure, and it even had a patriotic appeal. Vitus Bering, the man who in 1728 discovered the strait between Russia and America, was a Dane.

While in port, the *Odyssey* was less an intrepid ship and more of a floating warehouse. A metal gangway connected it to the pier and was watched at all hours by two members of the crew. It was important that nothing extra be allowed to get on board (drugs, for example, or tanks) but that all the proper things (maps, food, cigarettes) did. Most important of all was the iron ore. It had to be loaded as efficiently as possible in the ship's seven deep cargo holds, but also as evenly as possible: nothing can take a ship down faster than its cargo, improperly loaded. There was also the depth of the water to keep in mind. Fully loaded, the *Odyssey* would have a draft—the plumb distance from the waterline to the keel—of 43 feet; at the pier, the water at low tide was 42 feet. Thus the ship had to load up at high tide and then leave.

Petersen spent two days in Murmansk and then flew back to Copenhagen. The responsibility for loading the *Odyssey* fell on its chief mate, Vadim Zakharchenko. He was a short, broad-shouldered man with red hair and freckles; in his dark jumpsuit, he resembled a small bear. A native of the old port city of Odessa, he spoke Russian with a surprising Yiddish lilt—a legacy, he said, of his many Jewish classmates. On the early morning of July 9, the *Odyssey*'s last day in port, he was in a foul mood. The stevedores had told him that they weren't going to get to 65,000 tons of iron ore in time. In fact, Zakharchenko reported to Igor Shkrebko, the captain of the *Odyssey,* "They say we'll be lucky to reach sixty-four." At current shipping prices, a thousand fewer tons would put Nordic Bulk down between $20,000 and $30,000: an inauspicious start to the trip.

The captain was a tall, thin man, still youthful in his midforties, with curly, graying hair and black eyes. During the stay at Murmansk, his young wife had come up from their hometown of Sevastopol to visit; while most of the crew stayed on board, the Shkrebkos had walked around town and taken lots of photographs. In any case, cargo loading was the chief mate's job. "Akh!" Zakharchenko finally said. "They'll throw what they throw!"

For the rest of the morning, he scampered among the cranes and dockworkers, balancing two conflicting imperatives: that the cranes load the ship at record speed and that the hills of iron ore remain evenly distributed throughout the holds. The tall yellow cranes worked with urgency, picking up 6 or 7 tons of ore from the mounds piled on the dock, swinging over the cargo holds, then releasing the ore with a swoosh. As a light rain began to fall, Zakharchenko several times climbed down a rope ladder to the lee side of the ship to check how far it had descended into the water. Each centimeter represented 67 tons; incredibly, this was the only way to measure how much ore the *Odyssey* had taken on.

High tide was at noon, and the ship could not stay at the pier any longer. At eleven-thirty the cranes stopped loading, and fifteen minutes later all was done. According to an eyeball measurement of the ship's displacement, taken by both Zakharchenko and a surveyor hired by the Russian company that was shipping the ore, and a somewhat hurried calculation of the water density in the harbor, the *Odyssey* was now filled with 67,519 tons of ore: 2,500 tons more than the target. The stevedores had underestimated themselves. Those stevedores now ran down to the dock and removed the ship's thick ropes from the bollards; then three small tugboats came alongside the *Odyssey,* two to push and one to pull the ship into the harbor. That night, as the sun dipped toward the horizon (though it would not set), we entered the Barents Sea. You could tell it was the sea because right away our ship, despite now weighing more than 80,000 tons, started listing from side to side atop the waves.

*

Ahead of us, to the north and to the east, the ice was melting. This was normal. At its maximum extent, in mid-March, the ice covers the entire Arctic Ocean and most of its marginal seas for about 15 million square kilometers, twice the land area of the continental United States. During its minimum extent, around mid-September, the ice cover traditionally shrinks to about half this size.

In recent years, it has been shrinking by much more than half. In September 2007, the ice shrank to 4.3 million square kilometers, the lowest extent in recorded history. In subsequent years, it reached its second-, third-, and fourth-lowest-ever extents. The thickness of the ice—more difficult to measure but also more telling—is also decreasing, from an average thickness of 12 feet in 1980 to half that two decades later. The primary cause of this decline is warmer air temperature in the Arctic, an area that has been more affected by global warming than any other place on Earth.

The estimates vary, but scientists agree that at some point in this century the minimum extent, at the end of the summer season, will reach zero. At that point you'll be able to cross the North Pole in a canoe. But it won't be just you and your canoe, because the resource grabs have already begun. Denmark and Canada are engaged in a territorial dispute over Hans Island, which a recent congressional research report describes as a "tiny, barren piece of rock" between Greenland and Canada's Ellesmere Island, because territorial claims will lead to resource rights. Similarly, Russia has filed a claim with the United Nations that the Lomonosov Ridge, which spans the Arctic underwater from the coast of Siberia to Ellesmere Island, gives Russia rights to the sea above it, including the North Pole. All this is being done in anticipation of a thaw. Oil companies, armed with new technology and lured by less menacing winter conditions, will be able to establish drilling platforms in latitudes that were previously off-limits, and shipping companies will be able to save time and money through the Arctic shortcut. Shell has already announced plans to begin drilling exploratory wells off northern Alaska. Last year, Rosneft, Russia's biggest oil company, signed a joint-venture agreement with ExxonMobil to proceed with oil exploration in the Kara Sea—once called Mare Glaciale, the "ice sea." Meanwhile, the *Odyssey*'s trip was a test case for the proposition that the Northern Sea Route, formerly known as the Northeast Passage, could be reliably traversed.

*

The water of the Barents was a handsome dark blue, the sky was clear, and the temperature outside, though gradually dropping, was a balmy 50 degrees. Captain Shkrebko set our heading east for the southern tip of the archipelago Novaya Zemlya; this put the ship at a better angle to the waves, and it stopped rocking. We were proceeding at an unimpressive speed, 13 knots, but then again we never stopped. Three bridge crews of two men each, an officer and an able seaman, carried out four-hour shifts throughout the day and night.

The *Odyssey* had a permanent complement of just twenty-three men. The senior officers — captain, chief engineer, chief mate — were Ukrainian, as were the electrician and the second engineer; the rest of the crew was Filipino. The Ukrainians spoke Russian among themselves, while the Filipinos spoke Tagalog. Across the cultures, they spoke a rudimentary marine English. Otherwise, their contacts were limited. In addition to this permanent crew, there was a Russian "ice adviser," or pilot, named Eduard Cherepanov, who had been sailing these waters for almost twenty years.

Relations aboard the *Odyssey* were hierarchical and traditional. The captain, a native of the old naval city of Sevastopol, was the absolute authority. He was therefore a little isolated socially from the crew and seemed grateful for the presence of Cherepanov, who had served as a captain and was therefore his social equal and, more important, not someone with whom discipline needed to be maintained.

The chief mate, Zakharchenko, occupied an ambiguous position. On the one hand, he was in charge of much of the day-to-day operation of the ship, and he was the only one on board who knew as much about ships and the sea as the captain. On the other hand, he was entirely at the mercy of the captain, not only on the ship but professionally: because the chief mate has no independent contact with the home office, the only way he'll ever get a captaincy is if he's actively promoted by his captain. Zakharchenko was a soft touch. He tried to present a stern face to the crew, but he stuttered when he was nervous, and when he wasn't nervous he couldn't help but make a joke of some kind. As he liked to say, "Am I from Odessa or not?"

The two most senior Filipino crew members were Felimon Recana, the second mate, and Eliseo Carpon, the third mate. Both men were in their fifties, almost a decade older than Captain

Shkrebko and Zakharchenko. The second mate was handsome and sarcastic, a born cynic; the third mate was gregarious and enthusiastic. I once saw him jump up and cry "Yes!" after winning a game of Spider Solitaire on the computer in the crew rec room.

Life on board the ship is mostly confined to the "accommodation," a yellow, five-story metal building that rises from the stern. The bridge is on the top floor; the bottom floor contains locker rooms for the men as they prepare to go on deck. The men's living quarters are spread through the second, third, and fourth floors. Each man has his own cabin, about the size of a college dorm room, with a small bathroom and shower. Everything is secured so that it doesn't go flying around the room during a storm. This battening down takes some getting used to. It's easy enough to understand why the mini-fridge is strapped to a hook in the wall and the back of the bathroom mirror has little compartments for your toothbrush and shaving cream, but it took me almost a week to realize that the drawers under my bed, which wouldn't open when I tried them, were not ornamental, as I'd decided, but just extremely sticky. I was able to move my clothes out of my desk.

Most of the men were on six-month contracts, with monthly pay ranging from $1,100 for the mess boys to around $10,000 for the captain and the chief engineer—pretty good money in the Philippines and Ukraine. The contract is the standard unit of experience in the trade; one says "my last contract" rather than "my last ship." A six-month contract may include as few as ten port calls and as many as several dozen. These reprieves are short, and growing ever shorter as improved port technology gets ships in and out faster, but the men are grateful for them and can recite the price of girls in many ports across the world. The crew members had all received phone calls from their crewing agency in early April and had taken over the ship from its previous crew at the Irish port of Aughinish in mid-May. So far they'd brought soybeans from Quebec to Hamburg and coal from Latvia to Antwerp. None of them had been through the Arctic before.

To be aboard a ship is to be constantly aware of everything that can go wrong. A ship can run into another ship—hard to believe when you look at how wide the ocean is, a little easier to believe once you consider that it takes the *Odyssey* almost two miles to come to a complete stop. A ship can be overtaken by pirates:

Captain Shkrebko narrowly escaped pirates in the Gulf of Aden in 2007 (he was saved when an American military helicopter responded to his distress call), while the *Odyssey*'s fourth engineer was on a ship that was hijacked off the coast of Kenya in late 2009 and held hostage for forty-three days. A ship can be compromised by its cargo, which may shift, forcing the ship off balance, or create other problems—Zakharchenko had with him an alarming color brochure called "How to Monitor Coal Cargoes from Indonesia," which warned that Indonesian coal had a tendency to catch on fire. The *Odyssey*'s electrician, Dmitry Yemalienenko, had a short cell-phone video of a ship listing very hard to starboard in the Black Sea; it was carrying plywood, which had shifted en route. "Then what happened?" I asked.

"It sank," Yemalienenko said.

Then there was the danger of running into something beneath the waterline. To avoid this, the ship carried a full set of hydrographic charts, most of them from the British Admiralty. But the charts are never complete, and the telex machine on the bridge kept up a steady patter of warnings. When we headed out into the Barents, there was a broken signal at 69°40'N, 32°09'E, a shipwreck at 69°52', 35°16', nighttime artillery fire at 70°15', 33°38', plus some fishing nets.

Finally, there is the ice. The books on the bridge of the *Odyssey*—arranged on shelves behind the navigation table, with little wood braces to keep them from falling out in heavy seas—were all in agreement on the subject of the ice. "It is very easy and extremely dangerous to underestimate the hardness of ice," *The Mariner's Handbook* cautioned. "Ice fields consisting of thick broken floes, especially those that bear signs of erosion by the sea on their upper surface, should be avoided . . . Do not enter ice if a longer but ice-free route is available." *The Guide to Navigating Through the Northern Sea Route,* published in English in 1996 by the Russian Ministry of Defense, put the matter more dramatically: "Any attempt at independent, at vessel's own risk, transiting the NSR, without possessing and using full information, and without using all means of support, is doomed to failure."

This seemed harsh. But the ice-strengthened cruise ship *Explorer* sank off the coast of Antarctica in 2007 after hitting ice. The shrimp trawler *BCM Atlantic* sank near Labrador after hitting ice in 2000. Were the *Odyssey* to start sinking, there was a free-fall lifeboat hanging three stories up and at a 45-degree angle from the stern, but Vadim Zakharchenko said he would rather drown; the boat is raised so far up that its impact against the water could knock out your teeth.

Seamen don't like to talk about the things that can go wrong at sea, but they love to talk about the things that go wrong on land. As we approached Novaya Zemlya, the Ukrainians started joking about radioactivity. The Soviets had turned Novaya Zemlya into a nuclear-testing site; while they were at it, they used the coast around it as a dumping ground for reactors from decommissioned nuclear submarines. The largest nuclear bomb in history, the Tsar Bomba, had been detonated here in 1961. "Chernobyl is nothing compared to this!" Vadim announced.

On the evening of July 11, we entered a thirty-mile-wide strait between the southern end of Novaya Zemlya and Vaygach Island, at the entrance to the Kara Sea. The southern portion of the Barents that we had just been through is open to warm Gulf Stream currents, and it's rarely frozen even in winter. The Kara Sea is a different story. For years, no one could penetrate it. In the 1590s the Dutch explorer Willem Barents was repeatedly foiled by the ice at the Kara Gates and decided at last to head north and seek a way around Novaya Zemlya. This was not a good idea. His ship became trapped in ice, and the crew was forced to abandon it and spend the winter on land. One evening in October, the sun set and did not come back up again for three months. The men battled cold, scurvy, and hungry polar bears. "In Nova Zembla," the chronicler of the journey wrote, "there groweth neither leaves nor grasse, nor any beasts that eate grasse or leaves live therein, but such beasts as eate fleshe, as bears and foxes." When the warm weather came, in June, the crew headed for the Russian mainland. Some survived; Barents died of scurvy on the way.

The failed Barents expedition took place during the late-sixteenth-century Dutch ascendancy on the seas. It followed failed English attempts to traverse the passage earlier in the century and preceded some failed Russian ones. To be fair to these early explorers, their boats were made of wood, their maps were wildly inaccurate, they didn't know what a vitamin was, and they had no satellites to help them navigate the ice. Instructions from the London-based Russia Company to its early employees were notably vague: "And when you come to Vaygach, we would have you to get sight of the maine land . . . which is over against the south part of the same island, and from thence, with Gods permission, to passe eastwards alongst the same coasts, keeping it alwayes in your sight . . . untill you come to the country of Cathay, or the dominion of that mightie emperour." This was the state of the art in 1580. The dream was to reach China and its untold riches. But after enough men had disappeared into the ice never to return, the Dutch and the English decided it would be easier to go to war with Spain and Portugal for the right to use the route around the south of Africa; the Arctic, for a while, was forgotten.

For the next 950 miles, the Russian mainland stretched upward into the Arctic, forcing us to head northeast through the Kara Sea. Only when we reached Cape Chelyuskin, at almost 78 degrees N the northernmost point in Asia, could we turn southeast. And the farther north we got, the colder it became. Out on deck, though the temperature was still above freezing, a chill northerly wind blew in our faces.

On the morning of July 13, we crossed the 75th parallel; we had passed by the Yamal Peninsula, home to most of Russia's natural gas, and the mighty Ob and Yenisey Rivers. In recent years these rivers have been discharging more fresh water into the Arctic seas, as warmer temperatures increase overall precipitation in the Arctic water basin. Scientists anticipate that there will soon be more soil in the water, as the permafrost layer underground melts and the riverbanks begin to slide down. The Kara Sea was clear and cool, the air temperature 39 degrees, the water temperature 41; not swimming weather, but nothing to make ice from, either.

Late in the morning, we entered a stretch of fog. We could see as far as the bow of the ship and not an inch farther. The captain turned on our foghorn. It emitted a deep, loud wail every two minutes, to let anyone in front of us know that we were coming. But the ice pilot thought this precaution was goofy. "We don't really need that thing, you know," he said to the captain. "There's no one else out here." He was right. That afternoon, we were in radio contact with the two ships that were joining us in our convoy through the Northern Sea Route; one was 150 miles ahead of us, the other 100 miles behind. That afternoon, too, on our radar, we saw the only other boats outside our convoy that we'd encounter on the Northern Sea Route: the *Geofizik* and the *Geolog Dmitriy*

Nalivkin—the ExxonMobil/Rosneft seismic expedition, searching for oil.

The next morning we finally saw it: ice. It floated in isolated islands along the water. The islands were 10 or 15 feet in diameter, with a layer of snow on top, which protruded from the water by about a foot; beneath the water, you could see the ice, a few feet down and widening toward the bottom before narrowing again, like a teapot. These ice floes were on their way out of this world: there were still two months left in the melting season, and already the floes looked the worse for wear. The water lapped at their corners. In the middle of some of the floes, little green pools, known as "melt ponds," had formed in the snow. Unlike the white snow cover, which reflected sunlight back into the atmosphere, the puddles absorbed it. The sunlight was slowly drilling a hole in the ice under the puddles; if it managed to create a hole all the way down to the water, the water would have a toehold inside the ice to begin its destructive work.

They were a strange sight, these islands of ice in the middle of the sea. The lookout on the Barents expedition, when he first encountered the ice, exclaimed that he saw swans. Our crew was equally amazed. Many of the younger men were immediately on deck with digital cameras and cell phones. Eliseo, the third mate, who'd been going to sea for twenty-five years, was especially moved. "My first time," he said.

A few hours later, we reached the rendezvous point with the icebreakers: the *Vaygach,* beige and black, and the *Yamal,* red and black. They were not as long as the *Nordic Odyssey,* but they were stouter and, with their nuclear-powered engines, significantly more forceful. They had shallow-angled bows that allowed them to climb atop ice and crush it with their weight. A shallow bow must have felt insufficiently aggressive to the builders of the *Yamal,* however, for they had painted on it a set of big red jaws.

A Norwegian tanker, the *Marilee,* its deck covered with a tangle of pipes by means of which it kept its various liquids separate, was also waiting for us at the rendezvous point, and in the middle of the afternoon a 570-foot Russian cargo ship, the *Kapitan Danilkin,* caught up with us as well. Off we went into the ice. We were now approaching the tip of the Taymyr Peninsula, Cape Chelyuskin, named for the explorer who reached this spot by land in 1742.

Halfway between Murmansk, to the west, and the Bering Strait, to the east, it was one of the most obscure places in the world; Severnaya Zemlya, a large archipelago just thirty miles north of the cape, was not discovered until 1913—the last major piece of undiscovered land on Earth. The ice we'd seen earlier was scattered and melting; this ice was thicker and packed closer together. We followed the *Yamal* at a distance of about half a mile; the *Vaygach* was behind us, followed by the *Marilee* and the *Kapitan Danilkin*. We were soon joined by a small red tugboat, the *Vengery*, which took its position directly behind the *Odyssey*.

Captain Shkrebko, who until this point had mostly been taking photos with an expensive camera, walking around in sneakers, and generally looking more like a club tennis pro than like a sea captain, was now fully engaged, giving minute instructions to Able Seaman Ronald Segovia, who was at the wheel. The captain and the ice pilot had both gotten up in the middle of the night at the first sight of the ice and were still up, twenty hours later. Their job was to maintain radio contact with the icebreaker ahead and help the young helmsman maneuver the ship in unfamiliar conditions. Shkrebko and Cherepanov also had to decide how fast to go. There was a booklet on the bridge, from the Central Marine Research and Design Institute, in St. Petersburg, indicating the proper speed for an ice-class vessel through varying thicknesses of ice; the thicker the ice, the slower the ship should travel, so as not to damage its hull. But determining the actual thickness of the ice was an inexact science, and the ice pilot's contribution was primarily a counsel to remain calm.

"Take it down to six?" the captain would ask the pilot as they looked at one of the booklets, referring to 6 knots, or about half speed.

"Eight is probably fine here," the ice pilot would say, and we'd go to 8.

I put on a winter coat and hat and walked to the bow. It was a cold day and overcast. About 20 feet above the water, I watched the ship smash into the ice. Even after getting worked over by the *Yamal*, some of the ice pieces were big, 6 or 7 feet thick and 30 or 40 feet across. But we were bigger. Sometimes the ice simply cracked in two as soon as we collided with it and then fell away to our port and starboard. At other times it remained intact, trying to stop us, sometimes climbing the bow as we pushed it backward.

Occasionally a large piece would seem to have some traction, but the *Odyssey* was just too strong. Eventually the ice floes slithered off to the side. After we'd made it through the first ice field, the captain went down to the bow, too, and looked over the side. "Not even a scratch," he reported. He did not go down there again.

Over the next few hours, and then over the next eight days, we saw an incredible variety of ice. Some of the bits were just a few feet across, some were hundreds of feet; some were gray and even black, covered in grime, the way snow gets in New York after a few days. Some of the ice floes bobbed up and down in our wake; others remained proudly immobile. A few times the ice was so thick, and the icebreaker broke it so cleanly, that it came up again on its side, looking like a giant slice of cake, with green and blue layers separated by thin lines of white. Sometimes a smashed ice floe would be submerged beneath the surface and then come up, the water rolling off its back as off a slowly rising whale.

It took the *Odyssey* nearly twenty-four hours to round Cape Chelyuskin and enter the Laptev Sea. The sun still hadn't set since we'd left Murmansk, and much of the time the skies were relatively clear. But the air temperature was now at freezing, and toward the middle of the afternoon, on July 15, it began to snow.

As the trip progressed, I found myself spending more time with the chief mate, Vadim. Of all the men on board, he seemed the most ambivalent about his job, and the most philosophical. "This sun-filled prison," he said of the bridge. "A wonderful people," he said of the Jews of Odessa. "They've all left. And I alone in that whole city to carry on their memory."

Vadim's mother was a schoolteacher and his father an electrical engineer on a ship in the Soviet merchant marine. Young Vadim worshiped and feared his father. "He would come home from sea and you could just feel the aggression in him," Vadim said. "Then after two weeks he'd go back to normal." Seamen were a privileged category of Soviet citizen in that they could travel abroad, and Vadim, too, wanted to travel. He got his wish. In more than twenty years at sea, he has worked on passenger ships, refrigerator ships (reefers), oil tankers, and all kinds of bulk carriers, or bulkers. He likes to talk about music, soccer, and citizens of Odessa who have become wealthy, but his favorite topic is how sick he is of the sea. "You think it's beautiful," he would say as the sun came out from behind a cloud and shone on the blue, clear water, lightly chopped by the wind. "I used to think it was beautiful, too. Now I can't even look at it."

Vadim has other regrets about his career. "I became chief mate too late," he told me. "I was thirty-five. At that age, some people are already captains." Vadim was a captain just once in his career. He had joined the crew of a Greek bulker in South Korea, which set out for Seattle to get yellow corn. Before the trip began, he had a bad dream: he was naked, and when he looked down he saw that he was a woman, not a man. A bad omen. A week into the trip, the captain said he had a pain in his side. By the morning he had died. Vadim was now acting captain of the ship, and he called the home office in Athens. "The Greeks asked me if I had a captain's license," he said. "If I'd had one, I think they would have told me to keep going. Imagine showing up in the U.S. with a body on board? I'd have spent weeks filling out paperwork. I'd probably still be there!" In the event, Vadim did not have a captain's license. The ship returned to South Korea, a new captain flew in, and Vadim went back to being chief mate.

Vadim has a lightning-quick mind for arithmetic and a fondness for record keeping. He has a folder on his laptop called "1,001 Songs," containing his favorite songs from all over the world, with not a single artist repeated. He keeps statistics, independently of the newspapers, for the Odessa soccer club, the Chernomortsi, and he sometimes has occasion when he's on land to send a correction to the papers when they've made a mistake. He has a file called "History," in which he lists every country he's ever visited, every major canal he's passed through, and every time he's crossed the equator. Vadim is forty-three, divorced, and has a daughter in college. He has kept a color-coded chart, month by month since 1993, of when he's been home in Odessa and when he's been at sea. The chart indicates that he's been at sea for twelve of his last twenty birthdays. In most of his photos from home, the chief mate is drunk.

On July 15, in the Laptev Sea, Vadim was in midsentence on the bridge when he suddenly stopped, walked over to a pair of binoculars, and looked through them north-northeast. "Iceberg," he said. I thought he was kidding. The third mate, Eliseo, had taken to saying *"Titanic"* to me every time we saw a more or less healthy piece of ice. But Vadim wasn't kidding. About 8 miles from us, well out of our way but within sight, a giant piece of ice sat regally in the water. It had most likely calved off one of the glaciers on Severnaya Zemlya. Vadim estimated that it was about 65 feet high and perhaps 300 feet long.

We continued on our way through the Laptev Sea. In September 2007, when the ice receded to what was then its all-time minimum, the Northern Sea Route was still very difficult to navigate, because a 300-mile belt of drift ice remained bunched up in the Laptev. But now the Laptev was nearly empty of ice.

Each day we received reports on weather and ice conditions in the Arctic, but aside from that our information was limited. We had no Internet access aboard the ship. The captain was able to send and receive e-mails from a computer on the bridge, and others were theoretically allowed to send e-mails from the same computer, with the captain printing out the replies and slipping them under your door, but none of the crew members seemed to avail themselves of this service. Contact with home was confined to the satellite phone in the ship's office, which charged 50 cents a minute. "It's hard without the Internet," Vadim said. "You don't know who got blown up, who got assassinated. A few years ago, I came home and it was months before I found out that Yeltsin had died!"

Some of the crew wanted news of their families and called home weekly; some did not. The second mate, Felimon, claimed that he never did. "If I call from sea and there is problem," he said, "and then I call from port—it is same problem. There is nothing I can do." Able Seaman Edison Vocal told a story about a friend from a previous contract. The friend had received word from home that his wife was seeing someone else. For several weeks he kept himself from calling—what was he going to do out at sea?—but finally he called. His daughter answered. Mommy had a guest over, she said, and couldn't come to the phone. Edison's friend became depressed. He stopped eating. Then he jumped overboard. The ship went back and found him, but that was the sort of thing that could happen if you called home.

The crew entertained themselves as best they could. At six P.M. each day, four of the Filipinos would play doubles Ping-Pong in the gym. The level of play was erratic. The mess boy, Reynaldo Dalinao, the youngest crew member, always tried to slam the ball, with mixed success. Ordinary Seaman Michael Arboleda, whose day job mostly consisted of washing the ship and who was tall and broad-shouldered and always wore a basketball jersey with his last name on it (his cousin is a professional basketball player in Manila), tended to hit the ball casually into the net, then laugh. The star player was the third mate, Eliseo, who used a strange, possibly experimental grip, placed the ball wherever he pleased, and waited to pull you out of position. This was unquestionably the most fun I ever saw the crew have.

Mealtimes were at seven A.M., noon, and five P.M. All the Filipino crew who weren't on shift would fill up the crew mess tables and eat and talk—though they rarely tarried over their meals, sometimes wandering over to the TV at the other end of the room if they had time to spare. The officers' dining room was different. Reynaldo, the mess boy, set out everyone's food—usually some form of cabbage soup, followed by fried beef and potatoes—and covered it with plastic wrap. The Ukrainians came and scarfed it down when they could, almost always alone. At most times of the day, you could find four or five plastic-covered meals sitting on the tables in the officers' dining room, growing cold.

In the evenings, a group would gather in the crew rec room to watch an American action film, though the Manny Pacquiao–Timothy Bradley fight, which ended in a controversial decision for Bradley, was also popular. The crew had learned about the decision in Hamburg, then bought a DVD in Antwerp. The third mate had seen the fight about six or seven times by his estimate, whereas Able Seaman Generoso Juan had seen it "every time," which he believed was closer to a dozen. The Ukrainians, meanwhile, all had their own laptops and tended to stay in their cabins in the evenings and watch Russian television serials that they had downloaded from the Internet before shipping out.

At 75 degrees latitude, the circumference of the Earth is a quarter what it is at the equator, which means that one's time zone changes every 269 miles. On the *Odyssey*, the ship's time was at the discretion of the captain, and in a sense it didn't much matter what the local time was, since the sun never set. But the captain figured that it would be better to adjust the clocks gradually, by increments of an hour, than to move them ahead eight hours when we finally reached the Bering Strait. And so one slowly lost a sense of what time it "actually" was somewhere else. The ship's time was the only time that mattered.

On July 17, as we passed north of the New Siberian Islands (where nineteenth-century explorers had found well-preserved mammoth remains) and entered the East Siberian Sea, our captain turned forty-five. Toward evening the Ukrainians and the ice pilot huddled in the captain's cabin for a small party. The captain opened some pickled vegetables he'd picked up in Murmansk, and Reynaldo brought up some bread and cheese from the galley. The ship's ban on alcohol was temporarily lifted, and we drank to the captain's health.

The captain came from a long line of captains. His grandfather had been a captain in the NKVD, and his father was a captain in the Soviet merchant marine. Young Igor began his career on a reefer off Antarctica, as Soviet fishermen harpooned their last whale before the international ban on whaling went into effect in 1986. After the Soviet Union fell apart, he'd remained with the old company. Those years were full of adventures, as Ukraine sold off its inheritance from the USSR. While still in the employ of a reefer company, Shkrebko towed an old warship to Turkey. A few months later, the authorities called him in: "They said, 'You sold a warship to Turkey.' I sold a warship to Turkey? 'I was hired to tug a ship to Turkey. Here's the contract. It went out of Sevastopol port in full view of your military, with all the proper papers and permissions and everything. I sold it?'" Eventually, there was nothing left to sell. Shkrebko began his first contract with an international shipping company in 2000. He was given his first command in 2006.

The other men had similar stories, which they told when the captain—who didn't necessarily like other people talking when he was talking—was distracted. They had been to hundreds of ports among them; they had met women from all over the world, had wooed them or paid them; they liked working for better money, for an international company, and with a mixed crew. (With an all-Soviet crew, there was always too much drinking: "At first, it's fine, but then guys start hitting each other in the face," Vadim said. "Then they wake up and can't remember who hit who in the face. It causes problems.") But they missed the Soviet merchant marine. The pay was worse but the friendships lasted longer. And the crews were coed. There was never any trouble finding companionship aboard the *Shota Rustaveli* or the *Maxim Gorky*.

Later that night I went down to the galley to get a drink of water. Someone was watching an adult movie in the crew rec room. On my way back up, I ran into Vadim coming out of the ship's office. The crew's satellite phone was in there, but whom would he have been calling? He was estranged from his ex-wife, and I knew he didn't have a steady girlfriend. The next day he admitted that he'd been calling a friend in Odessa to learn the latest scores of his beloved soccer team, the Chernomortsi.

The Russians, led by Vitus Bering, mapped the contours of the Northeast Passage, largely by land, in the 1730s and '40s, but it was only in 1878–1879 that anyone sailed the entire route, and it wasn't until the summer of 1932 that a ship, the icebreaker *Sibiryakov*, made the navigation in one season. Steel and coal, not high atmospheric concentrations of carbon dioxide, were what initially conquered the ice.

But what was happening now was unprecedented. When Mads Petersen, the cochairman of Nordic Bulk, first sent his cargo of iron ore from Norway through the Arctic in 2010, he did so in September, the month when the ice is at its minimum; he did that again in 2011. Never before had he sent a ship in July. But we were making decent time. And when the *Odyssey* came back through here in August, there would be less ice. When it came back again in September, there would be hardly any ice at all.

Yet Mads Petersen was the only person I talked to in the Arctic who believed in man-made global warming. The deputy head of Rosatomflot smiled when I asked him about it ("This stuff is cyclical"), and so did my friend Vadim, who thought that the theory of global warming was a Western hoax. Captain Shkrebko conceded that monsoons had grown stronger in recent years and that the tides and currents he encountered were not the ones indicated on the British Admiralty charts, but that was as far as he would go. And the ice pilot, Cherepanov, claimed to be especially tickled at the thought that the Earth was warming and the ice was melting. "So the UN did a study, huh?" he kept saying of the 2007 IPCC climate report, which I had made the mistake of citing. "Well, if the UN says it's true, it must be true." I gave Vadim a copy of a book I had brought with me about global warming, but I don't think his English was up to it, and it lay unread on the bridge until I took it back to my cabin.

Post-Soviets tend to be skeptical about global warming. But there are notable exceptions. Earlier this year, Vladimir Putin hosted a team of scientists from the Vostok Research Station, Russia's leading research station in Antarctica. In the 1980s researchers at Vostok were the first to extract an ice core covering a full glacial-interglacial cycle, which was crucial for confirming the hypothesis that carbon-dioxide levels and temperature are connected. So when President Putin asked Vladimir Lipenkov, from the Arctic and Antarctic Research Institute of St. Petersburg, whether the scientist really believed that human-made greenhouse gases were a significant factor in global climate, Lipenkov did not back down. "No one denies that," he said.

"No, no," Putin said. "There are experts who believe that the changes in the climate are unrelated to human activity, that human activity has just a minimal, tiny effect, within the margin of error."

Lipenkov's answer was categorical: "It is not within the margin of error. If you look at the last five hundred thousand years, according to the data from Vostok Station, it turns out that the level of carbon dioxide and the change in temperature are correlated; that is to say, they have always moved practically together. Right now, according to atmospheric measurements, the level of carbon dioxide in the atmosphere is significantly higher than at any time in the last five hundred thousand years."

In the East Siberian Sea, we encountered a different kind of ice from any we'd seen before. It was thicker and older, and, most impressive of all, it stretched north as far as the eye could see. The ice we'd encountered thus far was drifting along—it had become detached from the great polar ice pack—whereas the ice here was part of the pack, and it looked almost like land. It wasn't land, of course, and in fact it wasn't even stable; all the ice in the Arctic, since it lies atop the ocean, is subject to the currents of that ocean and is therefore always in motion. Because of the Transpolar Drift —which takes ice from the Russian side and past the pole, where it eventually floats by Greenland and into the Atlantic—the oldest ice in the Arctic is rarely more than ten years old.

But this system has been here continuously for millions of years, developing during that time a complete ecology, from the algae that bloom underneath the ice and the copepods that thrive on its edge, to the cod that eat them, to the seals that eat the cod, to the white bears, kings of the Arctic, whose great paws have widened over time so the bears can walk on ice that would seem too thin to support their weight. And seeing the ice that is at the center of this ecosystem, we smashed right into it.

We went slowly, at times very slowly. Looking out, you'd have thought we were in a snowfield—it was white in all directions, save for the black-and-red stern of the *Yamal*. It was now clear that we would make it through the ice. We were just too big not to. Yet at some point in the East Siberian Sea I began to hope that we would lose. Here was a landscape that we were simply causing to disappear. We carried 67,000 tons of iron ore. Add to this about 37,000 tons of coking coal, some limestone, and a lot of heat, and you could forge about 50,000 tons of steel—enough steel for three ships just like the *Odyssey*. And each of those ships would beget three more ships. We would breed ships like rabbits, and I wondered why. The owner of our ship, Mads Petersen, was in daily e-mail contact with our captain, and one time he called the satellite phone on the bridge to say hello. "Mr. Mads!" Captain Shkrebko exclaimed into the phone, and eventually passed the receiver to me. Petersen asked, Was it a great adventure? Yes, I said, it was a great adventure. And the ship, I added, was a powerful ship, which needed to fear no ice. "Yeah," Petersen agreed. "It's a lot of steel." He didn't yet know where we were docking in China, but he was pleased that the ship was on its way.

I found it impossible to dislike Mads, who had sent us on this journey as much out of curiosity as cupidity and who was not blithe about the circumstances. "On the one hand, yes, more shipping," he had said in Murmansk. "On the other hand—global warming." But I found now that I wanted him to fail, to be turned back, to have to address the next Arctic shipping conference he attended with a tale of woe. It was hard to see how this could happen. The only thing out here as big as us was the lonely iceberg we saw in the Laptev Sea.

On July 20, we reached Pevek, a small, sad port city in far northeast Russia, and parted ways with the cargo ship *Kapitan Danilkin* and the icebreaker *Yamal*. With the *Vaygach* in the lead, we continued eastward, now much closer to the shore, which was hilly, green, and snowy. This was Chukotka, land of the Chukchi. When the Swedish professor A. E. Nordenskiöld, the first man ever to complete a passage through the Northern Sea Route, met the Chukchi people in 1878, he found that they knew no Russian but could count to ten in English. They had more contact with the American whalers who had started coming through the Bering Strait than they did with the Russians. We were pretty far east.

I spent hours looking for polar bears. The bears were white, and the ice cover was white, so they weren't going to be easy to see. One night Vadim saw a walrus in the water and took a blurry photo of him. But bears do not typically hunt walrus, which are as big as bears and have huge, scary tusks. Bears prefer the smaller ringed seal. In recent years, as the ice has started melting earlier and receding faster, polar bears have been missing their chance to get on the ice for their summer hunting and have been forced inland, close to human beings, where they have a tendency to get shot.

I was beginning to count the days. I enjoyed not having to check my e-mail, but I wanted a beer and I was tired of the ship's loose-leaf tea: in the absence of a strainer, the leaves inevitably got into my mouth. Even the ice—so remarkable, so perishable—was starting to be a bit much. "Okay, we saw the ice, it was interesting" is how Vadim summed up the feeling. "But enough is enough." If it had been more difficult; if it had been more dangerous; if the passage were not already, in some ways, routine, perhaps we would have felt differently. I had lunch with Dima Yemalienenko, the electrician, and announced to him my view that we were just twelve days from China. (This turned out to be optimistic.) Dima shrugged. "I don't count the days until there's a month left on my contract," he said. "So we get to China, so what? It's just another city. When there's a month left on my contract, then I'll start counting." Vadim, for his part, admitted that when he got home from a contract he usually went to see a shrink.

After Pevek, there were just 300 miles until we emerged from the ice and rendezvoused with our sister ship, the *Nordic Orion*, but these were the slowest miles of all. It took us two long days to cover them, and the crew entered a kind of fugue state. There would be short periods of reprieve, and then the ice would appear before us again, looking like a jetty or even a coast. One morning I woke up at around five because it seemed to me that we had stopped. I went up to the bridge, and sure enough, we were trapped amid several large ice floes. The *Vaygach* had turned too sharply and we hadn't been able to follow. Vadim and the ice pilot and the captain were all on the bridge; they looked exhausted but also, somehow, relieved. One of the worst things that could happen to a seaman in the Arctic—we were technically beset—had just happened, but it wasn't so bad. Not far from here, in 1879, the American *Jeannette* expedition, which sought to reach the North Pole, became trapped in the ice. It then drifted northwest on the ice for a year and a half before finally being crushed: "It looked like a staved-in barrel," one witness said. The crew of thirty-two men managed to get off the ship and onto the ice, with three small boats and some provisions, and then made their way to the Siberian mainland, but one of the boats sank, while the two others became separated, and only thirteen crew members survived. This would not happen to the *Odyssey*. The *Vaygach* turned around to extricate us, but we kept our 20-foot propeller going, and eventually our immense mass got the better of the ice, which slipped off to the side. I wondered what the crew of the *Jeannette* would have made of us.

Late in the evening on July 21, two days from the Bering Strait, there was a radio message from the *Vaygach* that I didn't catch. The ice pilot was on the bridge, and he moved quickly to pick up a pair of binoculars. He said, "Bear."

At first, we couldn't see it. Then there it was: a small bear, not a cub but not fully grown either, about the size of a very large dog, and a little more beige than I'd expected. The creature was running along the ice, occasionally falling into the little ponds that formed in it, then getting back out again and running some more. It was at the most vulnerable age for a bear, weaned off its mother but not fully proficient at hunting. It was not yet fat.

Once in a while, it turned to face the *Odyssey* and opened its jaws wide for a roar. We couldn't hear it from where we were—especially not over the sound of our own engine—but it was definitely roaring at us. And it was running away.

At noon the next day, the *Odyssey* finally emerged from the ice. Waiting for us, on schedule, was the *Nordic Orion*, which was on its way to Murmansk to pick up iron ore and return with it through the Northern Sea Route to China. Also waiting was a Swedish oil tanker headed for Finland and a Chinese ship, the *Xuelong*, which was on a scientific expedition into the Arctic. It would be the first Chinese trip through the Northeast Passage, and it would raise fears of Chinese encroachment on the Arctic. The vessel itself was a Ukrainian-built cargo ship.

The *Vaygach* sent a small motorboat to ferry Cherepanov, the ice pilot, aboard the *Orion*, and then the *Odyssey* continued on its way. Toward evening we ran into a school of whales. They'd come up, spray water into the air, and then, with a flash of their big black tails, dive down again. It was a joy to watch. We saw probably fifty whales. American whalers had first gone through the Bering Strait and then east into these waters in the latter half of the nineteenth century, but it seems they didn't get them all.

The end of the ice and the sendoff from the whales made it feel as though we had bid farewell to the Arctic, but the Arctic had not yet bid farewell to us. Early on the morning of July 23, we saw what looked like land due east. This would have to be Alaska. But Alaska was more than 100 miles away—too far to see. "It's not Alaska," Vadim said. It was a mirage. The water was still cold, but the air was considerably warmer, and the result was a "superior mirage": we saw the dark line of the horizon twice, both where it actually was and at a phantom place above it. The mind interpreted the top image as land. This kind of mirage can happen anywhere but is particularly common in polar regions. The mirage was not something you could look away from, then look at again to find that it was gone. It was in its way a physical fact, and it kept up for hours. We never did see Alaska.

Around midmorning we reached the easternmost edge of Russia, which is also the easternmost edge of the Eurasian landmass: Cape Dezhnev. It is a sheer rock cliff, as dramatic and definitive as Cape St. Vincent, in Portugal, the southwesternmost point of Eurasia. In 1728 Vitus Bering had come through the strait from the south, rounded this cape, and then, running into ice a few miles farther along, decided to turn back. At the time, because he didn't continue to St. Petersburg, some people didn't believe him that there was a Northeast Passage. But he was right.

And so to China. We had, it seemed, been through so much, and yet we were only halfway there, still more than 3,500 miles from our destination; at our average sea speed, the remainder of our journey would take between eleven and twelve days.

We set our course southwest and turned the ship back on autopilot. Life returned to its pre-Arctic routines. When a ship is in port, it gets scratched and scuffed in a hundred different ways. It had been too cold and wet in the Arctic to do anything about the damage, but now the crew could begin repainting the winches and windlasses and greasing the chains that the salt water and the air had begun to rust. A day south of the Bering Strait, the crew saw the sun set for the first time in three weeks. It didn't go very far that first night and continued to project a dim, hazy light over the ocean, but the next night was as dark as any. The bridge crew started drawing a heavy blue curtain across the bridge to separate the illuminated section from the front, where the lookouts needed total darkness to see into the night.

The crew experienced boredom. What is boredom? Boredom is staring for hours at the smooth, mirrorlike water, hoping to catch a glimpse of something, anything. Boredom is deciding to create a tea strainer from a soda can, going down to the galley, cutting a can in half, poking holes in the bottom with a knife, and then cutting one's finger, pretty badly, on the aluminum. Boredom is not just showing up exactly on time for the nightly Ping-Pong tournament but holding a clandestine practice session during the afternoon. Less productively, boredom is playing Spider Solitaire on the computer in the rec room. Boredom is watching other people play Spider Solitaire in the rec room. The ship's champion was Vadim. He played on the third, most difficult level, and he won a quarter of his games. But he took no joy in it. "Motherfucker," he could be heard muttering at the computer. "Motherfucker."

As the days stretched on, people became grumpier. Discipline relaxed. Vadim may have stopped either showering or doing his laundry, because there was a slightly sour smell wafting from him. He also complained that his feet hurt. During a test of the emergency generator, Dima, the electrician, accidentally cut off all the electricity to the bridge, causing most of the instruments to shut down and every possible alarm on the bridge to sound. There was an immense racket, matched only by the yelling of the captain at the electrician, who yelled right back.

One morning I went up to the bridge at around six to find Vadim sitting with Able Seaman Generoso Juan watching American music videos on the chief mate's laptop. Vadim was delighted to see me. "Do you know this band?" he said. "It's called Blink 182. They play a form of music called 'punk rock.'" He proceeded to DJ a series of songs about Odessa, including the Bee Gees' "Odessa": "I lost a ship in the Baltic sea. I'm on an iceberg running free."

Mads Petersen still had not informed the captain of our destination in China, and the men discussed which port they'd prefer. Shanghai was the favorite—the city wasn't too far from port, and the girls were friendly—but it was unlikely we'd be going to southern China with iron ore, given that steel was mostly manufactured in the north. Maybe it wouldn't much matter where we ended up. Chinese ports are busy, and if the time in port is too short no one would get off anyway. Some of the men said they wouldn't go ashore even if there was time. It was expensive and possibly dangerous. Ordinary Seaman Alvin Piamonte said the Mafia had taken root in China, and he wasn't going ashore unless he had two or three guys with him, which could be impossible to arrange, given everyone's schedules. The ports the men most loved—the ones in Brazil, Australia, Vietnam—were friendly, warm, and relaxed. They used to like American ports, but after 2001, as part of the Global War on Terror, the United States abrogated centuries of international practice by severely restricting foreign seafarers' ability to go ashore. The men of the *Odyssey* always became agitated when discussing this. The only country as restrictive as the United States, they said, was Saudi Arabia. In the words of the second mate, "It has taken the little happiness we had and made it less."

The only way to cheer the men at such points was to remind them of Bangkok. In Bangkok, as soon as you arrive, a boat comes alongside and disgorges a portable bar, a restaurant, and many friendly young women. If you pay in advance, a woman will move into your cabin for several days, sleep with you, and get up in the morning and iron your shirts—all for about thirty dollars a day. In some ports the authorities turn a blind eye to this sort of thing. In Bangkok, according to Vadim, if you try to kick the party off your ship, your cargo simply won't get unloaded. For this reason, seamen love Bangkok.

In the last days of July, we passed by the disputed southern Kuril Islands, off the northern tip of Japan, and then we entered the Tsugaru Strait. After weeks of silence, the radar screen bloomed with hundreds of ships, of all different sizes, heading in all sorts of directions: container ships, the rectangular blocks stacked high on their decks like Legos; oil tankers, the pipes tangled on their decks like snakes; and small fishing boats, looking for tuna.

At last Mads Petersen informed us of our destination: a new port in northern China called Huanghua, 140 miles southeast of Beijing. Mads said that the port's maximum draft was 42 feet, and at first this caused consternation. "I did all the calculations," Vadim told the captain heatedly, "and even if the bilges are empty, and we've burned seven hundred tons of fuel, we're still at forty-three!"

"Stop yelling," the captain snapped.

Vadim became quiet. "Was I yelling?" he asked. The captain nodded.

But the crisis soon passed; Huanghua Port was expanding, and the authorities told us that 43 feet would be no problem. On the other hand, a port this new could hardly be expected to have much infrastructure for entertaining seamen or even much of a town. The men were disappointed but not surprised, and the second mate even offered the hypothesis that because the port was new the girls might be even cheaper—twenty dollars, he said. On the evening of August 4, we arrived at an anchor spot in the Bo Hai Gulf, 25 miles from the port, and, with a tremendous noise, dropped our 7-ton steel anchor. We were three days behind schedule, which, considering the unpredictability of the route, wasn't bad.

For four days we sat at anchor with nothing to do. The Bo Hai Gulf is less a sea than an oil-and-gas field with some salt water on it; not far from us, drilling platforms burned excess natural gas in the air. The only marine life that seemed to flourish in so dirty a sea was jellyfish, and we watched them float by our ship, hour after hour. By this point, we were out of flour and sugar. On the third day at anchor, we broke our last Ping-Pong ball. The crew had no maps, no friends, no guides to the city they were about to enter, and no way of getting them. All they knew was that the Chinese authorities had sent a very strict checklist of things that must not be aboard the ship when it came into port, including bugs. The ship had been entirely bug-free until entering the Bo Hai Gulf, which was in fact quite buggy. "It's *their* bugs!" the captain protested. Nonetheless, each day the crew would sweep the upper platforms, and Michael Arboleda would stalk around the corridors of the accommodation with a fly swatter, killing everything in sight.

On the morning of August 9, we were cleared to enter the port. It was hard at first to grasp how big it was. The Bo Hai Gulf in general—and this port in particular—was shallow, so the Chinese were dredging. By picking sand up from the bottom and moving it elsewhere, they had managed to make a canal that a ship like the *Odyssey* could travel through with room to spare. To protect the canal, they had constructed miles of breakwater. And still they were reclaiming land from the water, constructing a new pier several miles into the harbor. *"Molodtsi,"* the captain said. "Bravo." What seemed from a distance like the outlines of a town was in fact an array of warehouses, processing plants, and cranes. Later I read that during the reconstruction of the port, large bribes had been paid to the port company's chairman, Huang Jianhua. A court had sentenced him to death. It was an impressive port.

Two tugboats steered us to our pier next to a row of big red cranes. Vadim gave the order to open our cargo holds, and we all looked inside: the iron ore was there just as we'd left it in Murmansk, black, heavy, unshifted, and dry. We lowered our gangway to the pier; Michael and Alvin became security guards; and then we waited. The first person to visit us was our agent in the port, a tall young man who spoke halting English with a slight British accent. The men threw themselves upon him. They had gone on and on about the girls they were going to screw, for between twenty and fifty dollars, but now all they wanted was SIM cards for their phones so they could call home and Internet cards for their computers so they could Skype. Dima came onto the deck with his laptop to see if he could catch a free Wi-Fi signal, but there was nothing; he'd have to wait and pay.

The surveyors were next. There were three of them, all well dressed, thin, and friendly, and wearing what looked like expensive designer eyeglasses. They didn't speak much English, but they were shepherded into the ship's office and someone went to look for Vadim.

The last few days of the trip had seemed really to wear on Vadim. In addition to his smell, he looked tired and growled more than usual at Spider Solitaire; because his feet hurt, he'd started breaking his own rule against open-toed footwear on the bridge and wore sandals. Now, after making the Chinese surveyors wait, he tromped into the ship's office. He wore a white jumpsuit, its top five buttons unbuttoned so that his chest and a gold chain could be seen. He looked as if he hadn't slept, shaved, or showered in weeks. He looked angry. But I had stood with him that morning as we pulled into port and he recited the various differing qualities of ports worldwide, and I knew that this was the part of the trip he most enjoyed. I even wondered if he'd been preparing for this moment, like a great actor preparing for a part. The Chinese surveyors, who looked as if they all had degrees in mathematics, must have been frightened at the sight of him, and also relieved. This creature was unlikely to be able to read, much less out-math them.

Vadim then proceeded to get the better of the surveyors in at least three ways. First, after boarding a small boat and traveling around the perimeter of the ship, he bullied the youngest of them into accepting all his readings of the depth of the draft. "Thirteen twenty-three?" the surveyor would offer, and Vadim would snap, "Thirteen twenty-six! Absolutely!" I thought the surveyor would be offended by this, but he quickly grew accustomed to Vadim and laughed at everything he said. When it came time to measure the water density, Vadim dropped the hydrometer down to the very bottom, where the density would be greatest. As all this was going on, one of the younger crew members was walking around with another surveyor measuring the water in the bilge tanks. The less water he measured in the tanks, the more cargo we had, and the young crew member had been instructed by Vadim in the proper technique of bilge measurement. "Was I born in Odessa or not?" Vadim said.

After all the numbers were added up and multiplied, it turned out that he'd gone too far: we now had 200 tons more iron ore than when we left Murmansk. Vadim slapped his forehead and explained to the surveyors that he'd suspected the water-density readings had been off in the port of origin. Would the surveyors mind just signing for the lower, original number? The surveyors didn't mind. What were a few hundred tons of iron ore when you were receiving 50 million tons every month? China was going to swallow our little shipment and demand much more.

There were a few more formalities to take care of, and in the meantime some port traders came by and offered SIM cards and other small favors. There was no question of any girls coming on board, and there would hardly be time for a shore visit. The ship had never felt more like a prison. How long would it even be in port? The cranes were very large. The cargo holds were open. In the next two months, the *Odyssey* would go back to Murmansk and then back to China, then travel across the Pacific to Vancouver to pick up a load of coal, which it would take back to Hamburg via the Arctic route. Mads Petersen would meet the ship again in Hamburg in late November. "She looks basically the same as when I saw her last time," he would tell me. "I was actually a bit surprised that the effects were not greater." During the summer of 2012, the Arctic ice would set a record for melting, while the ships would set a record for cargo taken through the route. But that was in the future. For now, toward evening, exactly a month after we'd left Murmansk, one of the cranes swooped down from above, like an enormous red hawk, took the first pile of Russian iron ore, and deposited it on the Chinese pier.

STEVEN WEINBERG

The Crisis of Big Science

FROM *The New York Review of Books*

LAST YEAR PHYSICISTS commemorated the centennial of
the discovery of the atomic nucleus. In experiments carried out
in Ernest Rutherford's laboratory at Manchester in 1911, a beam
of electrically charged particles from the radioactive decay of radium was directed at a thin gold foil. It was generally believed at
the time that the mass of an atom was spread out evenly, like a
pudding. In that case, the heavy charged particles from radium
should have passed through the gold foil with very little deflec-
tion. To Rutherford's surprise, some of these particles bounced
nearly straight back from the foil, showing that they were being
repelled by something small and heavy within gold atoms. Rutherford identified this as the nucleus of the atom, around which
electrons revolve like planets around the sun.

This was great science but not what one would call big science.
Rutherford's experimental team consisted of one postdoc and one
undergraduate. Their work was supported by a grant of just £70
from the Royal Society of London. The most expensive thing used
in the experiment was the sample of radium, but Rutherford did
not have to pay for it—the radium was on loan from the Austrian
Academy of Sciences.

Nuclear physics soon got bigger. The electrically charged particles from radium in Rutherford's experiment did not have enough
energy to penetrate the electrical repulsion of the gold nucleus
and get into the nucleus itself. To break into nuclei and learn what
they are, physicists in the 1930s invented cyclotrons and other machines that would accelerate charged particles to higher energies.

The late Maurice Goldhaber, former director of Brookhaven Laboratory, once reminisced: "The first to disintegrate a nucleus was Rutherford, and there is a picture of him holding the apparatus in his lap. I then always remember the later picture when one of the famous cyclotrons was built at Berkeley, and all of the people were sitting in the lap of the cyclotron."

1.

After World War II, new accelerators were built, but now with a different purpose. In observations of cosmic rays, physicists had found a few varieties of elementary particles different from any that exist in ordinary atoms. To study this new kind of matter, it was necessary to create these particles artificially in large numbers. For this, physicists had to accelerate beams of ordinary particles like protons—the nuclei of hydrogen atoms—to higher energy, so that when the protons hit atoms in a stationary target their energy could be transmuted into the masses of particles of new types. It was not a matter of setting records for the highest-energy accelerators or even of collecting more and more exotic species of particles, like orchids. The point of building these accelerators was, by creating new kinds of matter, to learn the laws of nature that govern all forms of matter. Though many physicists preferred small-scale experiments in the style of Rutherford, the logic of discovery forced physics to become big.

In 1959 I joined the Radiation Laboratory at Berkeley as a postdoc. Berkeley then had the world's most powerful accelerator, the Bevatron, which occupied the whole of a large building in the hills above the campus. The Bevatron had been built specifically to accelerate protons to energies high enough to create antiprotons, and to no one's surprise antiprotons were created. What was surprising was that hundreds of types of new, highly unstable particles were also created. There were so many of these new types of particles that they could hardly all be elementary, and we began to doubt whether we even knew what was meant by a particle being elementary. It was all very confusing and exciting.

After a decade of work at the Bevatron, it became clear that to make sense of what was being discovered, a new generation of higher-energy accelerators would be needed. These new accelerators would be too big to fit into a laboratory in the Berkeley hills.

Many of them would also be too big as institutions to be run by any single university. But if this was a crisis for Berkeley, it wasn't a crisis for physics. New accelerators were built, at Fermilab outside Chicago, at CERN near Geneva, and at other laboratories in the United States and Europe. They were too large to fit into buildings, but had now become features of the landscape. The new accelerator at Fermilab was 4 miles in circumference and was accompanied by a herd of bison, grazing on the restored Illinois prairie.

By the mid-1970s the work of experimentalists at these laboratories, and of theorists using the data that were gathered, had led us to a comprehensive and now well-verified theory of particles and forces, called the Standard Model. In this theory, there are several kinds of elementary particles. There are strongly interacting quarks, which make up the protons and neutrons inside atomic nuclei as well as most of the new particles discovered in the 1950s and 1960s. There are more weakly interacting particles called leptons, of which the prototype is the electron.

There are also "force carrier" particles that move between quarks and leptons to produce various forces. These include (1) photons, the particles of light responsible for electromagnetic forces; (2) closely related particles called W and Z bosons, which are responsible for the weak nuclear forces that allow quarks or leptons of one species to change into a different species—for instance, allowing negatively charged "down quarks" to turn into positively charged "up quarks" when carbon-14 decays into nitrogen-14 (it is this gradual decay that enables carbon dating); and (3) massless gluons that produce the strong nuclear forces that hold quarks together inside protons and neutrons.

Successful as the Standard Model has been, it is clearly not the end of the story. For one thing, the masses of the quarks and leptons in this theory have so far had to be derived from experiment rather than deduced from some fundamental principle. We have been looking at the list of these masses for decades now, feeling that we ought to understand them, but without making any sense of them. It has been as if we were trying to read an inscription in a forgotten language, like Linear A. Also, some important things are not included in the Standard Model, such as gravitation and the dark matter that astronomers tell us makes up five-sixths of the matter of the universe.

So now we are waiting for results from a new accelerator at CERN that we hope will let us make the next step beyond the Standard Model. This is the Large Hadron Collider, or LHC. It is an underground ring 17 miles in circumference crossing the border between Switzerland and France. In it two beams of protons are accelerated in opposite directions to energies that will eventually reach 7 TeV in each beam, that is, about 7,500 times the energy in the mass of a proton. The beams are made to collide at several stations around the ring, where detectors with the mass of World War II cruisers sort out the various particles created in these collisions.

Some of the new things to be discovered at the LHC have long been expected. The part of the Standard Model that unites the weak and electromagnetic forces, presented in 1967–1968, is based on an exact symmetry between these forces. The W and Z particles that carry the weak nuclear forces and the photons that carry electromagnetic forces all appear in the equations of the theory as massless particles. But while photons really are massless, the W and Z are actually quite heavy. Therefore, it was necessary to suppose that this symmetry between the electromagnetic and weak interactions is "broken"—that is, though it is an exact property of the equations of the theory, it is not apparent in observed particles and forces.

The original and still the simplest theory of how the electroweak symmetry is broken, the one proposed in 1967–1968, involves four new fields that pervade the universe. A bundle of the energy of one of these fields would show up in nature as a massive, unstable, electrically neutral particle that came to be called the Higgs boson.* All the properties of the Higgs boson except its mass

---

* In his recent book *The Infinity Puzzle* (Basic Books, 2011), Frank Close points out that a mistake of mine was in part responsible for the term "Higgs boson." In my 1967 paper on the unification of weak and electromagnetic forces, I cited 1964 work by Peter Higgs and two other sets of theorists. This was because they had all explored the mathematics of symmetry-breaking in general theories with force-carrying particles, though they did not apply it to weak and electromagnetic forces. As known since 1961, a typical consequence of theories of symmetry-breaking is the appearance of new particles, as a sort of debris. A specific particle of this general class was predicted in my 1967 paper; this is the Higgs boson now being sought at the LHC.

As to my responsibility for the name "Higgs boson," because of a mistake in reading the dates on these three earlier papers, I thought that the earliest was the one by Higgs, so in my 1967 paper I cited Higgs first and have done so since

are predicted by the 1967–1968 electroweak theory, but so far the particle has not been observed. This is why the LHC is looking for the Higgs—if found, it would confirm the simplest version of the electroweak theory. In December 2011 two groups reported hints that the Higgs boson has been created at the LHC, with a mass 133 times the mass of the proton, and signs of a Higgs boson with this mass have since then turned up in an analysis of older data from Fermilab. We will know by the end of 2012 whether the Higgs boson has really been seen.

The discovery of the Higgs boson would be a gratifying verification of present theory, but it will not point the way to a more comprehensive future theory. We can hope, as was the case with the Bevatron, that the most exciting thing to be discovered at the LHC will be something quite unexpected. Whatever it is, it's hard to see how it could take us all the way to a final theory, including gravitation. So in the next decade, physicists are probably going to ask their governments for support for whatever new and more powerful accelerator we then think will be needed.

2.

That is going to be a very hard sell. My pessimism comes partly from my experience in the 1980s and 1990s in trying to get funding for another large accelerator.

In the early 1980s the United States began plans for the Superconducting Super Collider, or SSC, which would accelerate protons to 20 TeV, three times the maximum energy that will be available at the CERN Large Hadron Collider. After a decade of work, the design was completed, a site was selected in Texas, land bought, and construction begun on a tunnel and on magnets to steer the protons.

Then in 1992 the House of Representatives canceled funding for the SSC. Funding was restored by a House-Senate conference committee, but the next year the same thing happened again, and

then. Other physicists apparently have followed my lead. But as Close points out, the earliest paper of the three I cited was actually the one by Robert Brout and François Englert. In extenuation of my mistake, I should note that Higgs and Brout and Englert did their work independently and at about the same time, as also did the third group (Gerald Guralnik, C. R. Hagen, and Tom Kibble). But the name "Higgs boson" seems to have stuck.

this time the House would not go along with the recommendation of the conference committee. After the expenditure of almost $2 billion and thousands of man-years, the SSC was dead.

One thing that killed the SSC was an undeserved reputation for overspending. There was even nonsense in the press about spending on potted plants for the corridors of the administration building. Projected costs did increase, but the main reason was that year by year, Congress never supplied sufficient funds to keep to the planned rate of spending. This stretched out the time and hence the cost to complete the project. Even so, the SSC met all technical challenges and could have been completed for about what has been spent on the LHC, and completed a decade earlier.

Spending for the SSC had become a target for a new class of congressmen elected in 1992. They were eager to show that they could cut what they saw as Texas pork, and they didn't feel that much was at stake. The cold war was over, and discoveries at the SSC were not going to produce anything of immediate practical importance. Physicists can point to technological spinoffs from high-energy physics, ranging from synchrotron radiation to the World Wide Web. For promoting invention, big science in this sense is the technological equivalent of war, and it doesn't kill anyone. But spinoffs can't be promised in advance.

What really motivates elementary-particle physicists is a sense of how the world is ordered—it is, they believe, a world governed by simple universal principles that we are capable of discovering. But not everyone feels the importance of this. During the debate over the SSC, I was on Larry King's radio show with a congressman who opposed it. He said that he wasn't against spending on science, but that we had to set priorities. I explained that the SSC was going to help us learn the laws of nature, and I asked if that didn't deserve a high priority. I remember every word of his answer. It was "No."

What does motivate legislators is the immediate economic interests of their constituents. Big laboratories bring jobs and money into their neighborhood, so they attract the active support of legislators from that state, and apathy or hostility from many other members of Congress. Before the Texas site was chosen, a senator told me that at that time there were a hundred senators in favor of the SSC, but that once the site was chosen the number would drop to two. He wasn't far wrong. We saw several members of Congress change their stand on the SSC after their states were eliminated as possible sites.

Another problem that bedeviled the SSC was competition for funds among scientists. Working scientists in all fields generally agreed that good science would be done at the SSC, but some felt that the money would be better spent on other fields of science, such as their own. It didn't help that the SSC was opposed by the president-elect of the American Physical Society, a solid-state physicist who thought the funds for the SSC would be better used in, say, solid-state physics. I took little pleasure from the observation that none of the funds saved by canceling the SSC went to other areas of science.

All these problems will emerge again when physicists go to their governments for the next accelerator beyond the LHC. But it will be worse, because the next accelerator will probably have to be an international collaboration. We saw recently how a project to build a laboratory for the development of controlled thermonuclear power, ITER, was nearly killed by the competition between France and Japan to be the laboratory's site.

There are things that can be done in fundamental physics without building a new generation of accelerators. We will go on looking for rare processes, like an extremely slow conjectured radioactive decay of protons. There is much to do in studying the properties of neutrinos. We get some useful information from astronomers. But I do not believe that we can make significant progress without also pushing back the frontier of high energy. So in the next decade we may see the search for the laws of nature slow to a halt, not to be resumed again in our lifetimes.

Funding is a problem for all fields of science. In the past decade, the National Science Foundation has seen the fraction of grant proposals that it can fund drop from 33 percent to 23 percent. But big science has the special problem that it can't easily be scaled down. It does no good to build an accelerator tunnel that only goes halfway around the circle.

3.

Astronomy has had a very different history from physics, but it has wound up with much the same problems. Astronomy became big science early, with substantial support from governments, because it was useful in a way that, until recently, physics was not.* Astronomy was used in the ancient world for geodesy, navigation, timekeeping, and making calendars, and in the form of astrology it was imagined to be useful for predicting the future. Governments established research institutes: the Museum of Hellenistic Alexandria; the House of Wisdom of ninth-century Baghdad; the great observatory in Samarkand built in the 1420s by Ulugh Beg; Uraniborg, Tycho Brahe's observatory, built on an island given by the king of Denmark for this purpose in 1576; the Greenwich Observatory in England; and later the U.S. Naval Observatory.

In the nineteenth century, rich private individuals began to spend generously on astronomy. The third Earl of Rosse used a huge telescope called Leviathan in his home observatory to discover that the nebulae now known as galaxies have spiral arms. In America observatories and telescopes were built carrying the names of donors such as Lick, Yerkes, and Hooker, and more recently Keck, Hobby, and Eberly.

But now astronomy faces tasks beyond the resources of individuals. We have had to send observatories into space, both to avoid the blurring of images caused by the Earth's atmosphere and to observe radiation at wavelengths that cannot penetrate the atmosphere. Cosmology has been revolutionized by satellite observatories such as the Cosmic Background Explorer, the Hubble Space Telescope, and the Wilkinson Microwave Anisotropy Probe, working in tandem with advanced ground-based observatories. We now know that the present phase of the Big Bang started 13.7 billion years ago. We also have good evidence that before that, there was a phase of exponentially fast expansion known as inflation.

But cosmology is in danger of becoming stuck, in much the same sense as elementary-particle physics has been stuck for decades. The discovery in 1998 that the expansion of the universe is now accelerating can be accommodated in various theories, but we don't have observations that would point to the right theory. The observations of microwave radiation left over from the early universe have confirmed the general idea of an early era of inflation but do not give detailed information about the physical proc-

* I have written more about this in "The Missions of Astronomy," *The New York Review of Books,* October 22, 2009.

esses involved in the expansion. New satellite observatories will be needed, but will they be funded?

The recent history of the James Webb Space Telescope, planned as the successor to Hubble, is disturbingly reminiscent of the history of the SSC. At the funding level requested by the Obama administration last year, the project would continue, but at a level that would not allow the telescope ever to be launched into orbit. In July the House Appropriations Committee voted to cancel the Webb telescope altogether. There were complaints about cost increases, but as was the case with the SSC, most of the increase came because year by year the project was not adequately funded. Funding for the telescope has recently been restored, but the prognosis for future funding is not bright. The project is no longer under the authority of NASA's Science Mission Directorate. The technical performance of the Webb project has been excellent, and billions have already been spent, but the same was true of the SSC and did not save it from cancellation.

Meanwhile, in the past few years funding has dropped for astrophysics at NASA. In 2010 the National Research Council carried out a survey of opportunities for astronomy in the next ten years, setting priorities for new observatories that would be based in space. The highest priorities went first to WFIRST, an infrared survey telescope; next to Explorer, a program of midsized observatories similar in scale to the Wilkinson Microwave Anisotropy Probe; then to LISA, a gravitational wave observatory; and finally to an international x-ray observatory. No funds are in the budget for any of these.

Some of the slack in big science is being taken up by Europe, as for instance with the LHC and a new microwave satellite observatory named Planck. But Europe has worse financial problems than the United States, and the European Union Commission is now considering the removal of large science projects from the EU budget.

Space-based astronomy has a special problem in the United States. NASA, the government agency responsible for this work, has always devoted more of its resources to manned space flight, which contributes little to science. All of the space-based observatories that have contributed so much to astronomy in recent years have been unmanned. The International Space Station was sold in

part as a scientific laboratory, but nothing of scientific importance
has come from it. Last year a cosmic ray observatory was carried
up to the space station (after NASA had tried to remove it from
the schedule for shuttle flights), and for the first time significant
science may be done on the space station, but astronauts will have
no part in its operation, and it could have been developed more
cheaply as an unmanned satellite.

The International Space Station was partly responsible for the
cancellation of the SSC. Both came up for a crucial vote in Congress in 1993. Because the space station would be managed from
Houston, both were seen as Texas projects. After promising active
support for the SSC, the Clinton administration decided in 1993
that it could support only one large technological project in Texas,
and it chose the space station. Members of Congress were hazy
about the difference. At a hearing before a House committee, I
heard a congressman say that he could see how the space station
would help us learn about the universe, but he couldn't understand that about the SSC. I could have cried. As I later wrote, the
space station had the great advantage that it cost about ten times
more than the SSC, so NASA could spread contracts for its development over many states. Perhaps if the SSC had cost more, it
would not have been canceled.

4.

Big science is in competition for government funds, not only with
manned space flight and with various programs of real science,
but also with many other things that we need government to do.
We don't spend enough on education to make becoming a teacher
an attractive career choice for our best college graduates. Our passenger rail lines and Internet services look increasingly poor compared with what one finds in Europe and East Asia. We don't have
enough patent inspectors to process new patent applications without endless delays. The overcrowding and understaffing in some
of our prisons amount to cruel and unusual punishment. We have
a shortage of judges, so civil suits take years to be heard.

The Securities and Exchange Commission, moreover, doesn't
have enough staff to win cases against the corporations it is charged
to regulate. There aren't enough drug rehabilitation centers to
treat addicts who want to be treated. We have fewer policemen

and firemen than before September 11. Many people in America cannot count on adequate medical care. And so on. In fact, many of these other responsibilities of government have been treated worse in the present Congress than science. All these problems will become more severe if current legislation forces an 8 percent sequestration—or reduction, in effect—of nonmilitary spending after this year.

We had better not try to defend science by attacking spending on these other needs. We would lose, and would deserve to lose. Some years ago I found myself at dinner with a member of the Appropriations Committee of the Texas House of Representatives. I was impressed when she spoke eloquently about the need to spend money to improve higher education in Texas. What professor at a state university wouldn't want to hear that? I naively asked what new source of revenue she would propose to tap. She answered, "Oh, no, I don't want to raise taxes. We can take the money from health care." This is not a position we should be in.

It seems to me that what is really needed is not more special pleading for one or another particular public good, but for all the people who care about these things to unite in restoring higher and more progressive tax rates, especially on investment income. I am not an economist, but I talk to economists, and I gather that dollar for dollar, government spending stimulates the economy more than tax cuts. It is simply a fallacy to say that we cannot afford increased government spending. But given the antitax mania that seems to be gripping the public, views like these are political poison. This is the real crisis, and not just for science.*

* This article is based on the inaugural lecture in the series On the Shoulders of Giants, of the World Science Festival in New York on June 4, 2011, and on a plenary lecture at the meeting of the American Astronomical Society in Austin on January 9, 2012.

GARETH COOK

Autism Inc.

FROM *The New York Times Magazine*

WHEN THORKIL SONNE and his wife, Annette, learned that their three-year-old son, Lars, had autism, they did what any parent who has faith in reason and research would do: they started reading. At first they were relieved that so much was written on the topic. "Then came sadness," Annette says. Lars would have difficulty navigating the social world, they learned, and might never be completely independent. The bleak accounts of autistic adults who had to rely on their parents made them fear the future.

What they read, however, didn't square with the Lars they came home to every day. He was a happy, curious boy, and as he grew, he amazed them with his quirky and astonishing abilities. If his parents threw out a date—December 20, 1997, say—he could name, almost instantly, the day of the week (Saturday). And, far more usefully for his family, who live near Copenhagen, Lars knew the train schedules of all of Denmark's major routes.

One day when Lars was seven, Thorkil Sonne was puttering around the house doing weekend chores while Lars sat on a wooden chair, hunched for hours over a sheet of paper, pencil in hand, sketching chubby rectangles and filling them with numerals in what seemed to represent a rough outline of Europe. The family had recently gone on a long car trip from Scotland to Germany, and Lars had passed the time in the back seat studying a road atlas. Sonne walked over to a low shelf in the living room, pulled out the atlas, and opened it up. The table of contents was presented as a map of the continent, with page numbers listed in boxes over the various countries (the fjords of Norway, pages

34–35; Ireland, pages 76–77). Thorkil returned to Lars's side. He slid a finger along the atlas, moving from box to box, comparing the source with his son's copy. Every number matched. Lars had reproduced the entire spread from memory, without an error. "I was stunned, absolutely," Sonne told me.

To his father, Lars seemed less defined by deficits than by his unusual skills. And those skills, like intense focus and careful execution, were exactly the ones that Sonne, who was the technical director at a spinoff of TDC, Denmark's largest telecommunications company, often looked for in his own employees. Sonne did not consider himself an entrepreneurial type, but watching Lars—and hearing similar stories from parents he met volunteering with an autism organization—he slowly conceived a business plan: many companies struggle to find workers who can perform specific, often tedious tasks like data entry or software testing; some autistic people would be exceptionally good at those tasks. So in 2003 Sonne quit his job, mortgaged the family's home, took a two-day accounting course, and started a company called Specialisterne, Danish for "the specialists," on the theory that given the right environment, an autistic adult could not just hold down a job but also be the best person for it.

For nearly a decade, the company has been modest in size—it employs thirty-five high-functioning autistic workers who are hired out as consultants, as they are called, to nineteen companies in Denmark—but it has grand ambitions. In Europe, Sonne is a minor celebrity who has met with Danish and Belgian royalty, and at the World Economic Forum meeting in Tianjin in September, he was named one of twenty-six winners of a global social entrepreneurship award. Specialisterne has inspired start-ups and has five of its own around the world. In the next few months, Sonne plans to move with his family to the United States, where the number of autistic adults—roughly 50,000 turn eighteen every year—as well as a large technology sector suggests a good market for expansion.

"He has made me think about this differently, that these individuals can be a part of our business and our plans," says Ernie Dianastasis, a managing director of CAI, an information-technology company that has agreed to work with Specialisterne to find jobs for autistic software testers in the United States.

For previously unemployable people—one recent study found that more than half of Americans with an autism diagnosis do not attend college or find jobs within two years of graduating from high school—Sonne's idea holds out the possibility of self-sufficiency. He has received countless letters of thanks and encouragement from the families of autistic people. One woman in Hawaii wrote Sonne asking if she could move her family to Denmark so that her unemployed autistic son could join the Specialisterne team.

I first met Sonne, who is fifty-two, in Delaware at a small conference he organized for parents and government officials who want to help him set up American operations over the coming year. He stood before them, sipping a cup of Dunkin' Donuts coffee, speaking enthusiastically of his "dandelion model": when dandelions pop up in a lawn, we call them weeds, he said, but the spring greens can also make a tasty salad. A similar thing can be said of autistic people—that apparent weaknesses (bluntness and obsessiveness, say) can also be marketable strengths (directness, attention to detail). "Every one of us has the power to decide," he said to the audience, "do we see a weed, or do we see an herb?"

It's an appealing metaphor, though perhaps a tougher sell in the United States, where you rarely see dandelion salad. It is also, of course, a little too simple. Over eight years of evaluating autistic adults, Sonne has discovered that only a small minority have the abilities Specialisterne is looking for and are able to navigate the unpredictable world of work well enough to keep a job. "We want to be a role model to inspire," Sonne told me later, "but we can only hire the ones that we believe can fill a valuable role in a consultancy like ours." In other words, he's not running a charity. It is Sonne's ultimate goal to change how "neurotypicals" see people with autism, and the best way to do that, he has decided, is to prove their value in the marketplace.

TDC, Thorkil Sonne's former employer, is Specialisterne's oldest customer. When I visited its headquarters in Copenhagen in June, it was obvious why the company finds it useful to engage autistic consultants. Whenever cell-phone makers introduce a new product, there are countless opportunities for glitches. The only way TDC can be sure of catching them is to load the software onto a phone and punch the phone keys over and over again, following a lengthy script of at least 200 instructions. The work is tedious, the information age equivalent of the assembly line, but also important and beyond the capacity of most people to perform well. "You will get bored, and then you will take shortcuts, and then it is worthless," explained Johnni Jensen, a system technician at TDC.

Steen Iversen, a Specialisterne consultant in blue jeans and a bright red polo shirt, showed me how he tackles the task. Iversen, who is fifty-two and has worked at TDC for four years, laid out several phones on a desk that also held his computer, two bananas, an apple, and lines of lime-green Post-it notes. He picked up a phone in one hand and demonstrated his technique, his thumb landing on the buttons in quick succession. But his real advantage is mental: he is exhaustive and relentless. When a script called for sending a "long text message," Iversen keyed in every character the phone was capable of; it crashed. Another time he found a flaw that could have disabled a phone's emergency dialing capability, a problem all previous testers had missed. I asked Iversen how he feels at moments like that, and he gently pumped both fists in the air with a shy smile. "I feel victorious," he said.

Over the years, Jensen has developed strategies for interacting with Iversen and the two other consultants he oversees. Trying to rush them inevitably backfires, he told me. "Sometimes I have to bite my tongue." Jensen feels protective of the consultants and tries to shield them from the usual stresses of office work, but he is emphatic that the arrangement has endured not because he pities them but because their work is excellent. When Iversen finds a bug, he can recall similar ones from years past, saving Jensen the time and frustration of researching the problem's history. And, Jensen says, the consultants are far more devoted to accuracy than neurotypical workers. Iversen has punched mobile-phone keys day after day, and not once has he cut a corner or even made a careless mistake.

Christian Andersen, another Specialisterne consultant, works at Lundbeck, a large pharmaceutical company. He compares records of patients who have experienced reactions to Lundbeck's drugs, making sure the paper records match the digital ones. Errors can creep in when the reports are entered into the company's database, and tiny mistakes could mean that potential health hazards would go undetected. So Andersen searches for anomalies, computer entry against written report, over and over, hour after hour, day after day.

Before Andersen arrived, his boss, Janne Kampmann, had a

hard time finding employees who could do the job well. Most people's minds wander as they go back and forth between documents, their eyes skimming the typos lurking there. Andersen, however, worked without interruption the morning I visited, attentive and silent until he lifted his head and, pointing to a sheet of paper, said to Kampmann, "Why do we have a fifty-seven instead of thirty milligrams?" Kampmann told me Andersen is one of the best quality-control people she's ever seen.

For years scientists underestimated the intelligence of autistic people, an error now being rectified. A team of Canadian scientists published a paper in 2007 showing that measures of intelligence vary wildly, depending on what test is used. When the researchers used the Wechsler scale, the historical standard in autism research, a third of children tested fell in the range of intellectual disability, and none had high intelligence, consistent with conventional wisdom. Yet on the Raven's Progressive Matrices, another respected IQ test, which does not rely on language ability, a majority of the same children scored at or above the middle range—and a third exhibited high intelligence. Other scientists have demonstrated that the autistic mind is superior at noticing details, distinguishing among sounds, and mentally rotating complex three-dimensional structures. In 2009 scientists at King's College London concluded that about a third of autistic males have "some form of outstanding ability."

This emerging understanding of autism may change attitudes toward autistic workers. But intelligence, even superior intelligence, isn't enough to get or keep a job. Modern office culture— with its unwritten rules of behavior, its fluid and socially demanding workspaces—can be hostile territory for autistic people, who do better in predictable environments and who tend to be clumsy at shaping their priorities around other people's requirements.

Most Specialisterne consultants work in the offices of the companies that use their services, but some need to operate out of Specialisterne's more forgiving workspace. Even those capable of working onsite sometimes get into trouble. In one case, the company was contacted by a medical-technology company that needed help testing new prescription-tracking software. This seemed a marvelous bit of luck, says Rune Oblom, Specialisterne's business manager, because there was a consultant on staff interested in illnesses. Everything was going fine until a medical team arrived to try out the software, and the consultant spent the entire morning recounting to them, in detail, the medical treatments that he, his mother, and the rest of his family had received over the years. Another consultant was assigned to finish the software-testing job. "I told him that the doctors were not very happy and felt he was a disturbing factor," Oblom says. "But he couldn't see it."

The consultant has since been moved to another company, where he has done well at his professional tasks but still misses social cues. In Denmark there is a tradition of bringing cake to the office on Fridays, and Oblom recently learned from the onsite supervisor that the consultant happily eats cake but has never volunteered to bring one himself. Then there was the time he tasted a coworker's cake and pronounced it terrible. Oblom told me that he plans to tell the consultant that he has to bring in cake now and then—and he will do it, Oblom predicts, without understanding the reason—but he's not going to encourage the consultant to be more polite. The concept of socially mandated dishonesty would mystify him, Oblom said, so the other employees will just have to deal with it.

Specialisterne tries to anticipate, or at least mitigate, conflicts by assigning every consultant to a neurotypical coach. The coach checks in with the consultants regularly, monitoring their emotional well-being and helping them navigate the social landscape of the office. Henrik Thomsen, a jolly man who runs Specialisterne in Denmark while Sonne works on international expansion, told me about one consultant who is fascinated by train schedules. Severe storms can disrupt the trains around Copenhagen, and if the consultant's train was delayed, he would start the day with a tour of his colleagues at the Specialisterne office, telling each how the commute played out, station by station. Sometimes another consultant would get annoyed and tell him to "cut the crap," Thomsen says, "and then the real fun would begin." So now Thomsen listens to the radio as he drives in, taking mental note of potential delays. When Thomsen arrives at work, he invites the consultant into his office first thing, listens to the day's commuting story, and then asks him to please get to work.

Specialisterne's headquarters occupy part of a three-story complex in a Copenhagen suburb. Sonne showed me around the building: in addition to the consulting business, there is a nonprofit focused on spreading the Specialisterne business model and a small school for people on the autism spectrum in their late teens and early twenties. In the largest room, boxes of Legos are stacked against one wall, and a pair of long, waist-high tables for Lego activities occupies the center, under a string of halogen lights.

When Sonne started the company, one of his biggest challenges was determining who would be able to thrive as a tech consultant in an office environment. A traditional interview was clearly not going to do the trick, and he had to think of other ways to identify marketable strengths in people who have difficulty communicating.

Lars had always enjoyed Legos, and in talking to other parents, Sonne heard stories about how the toy bricks brought out remarkable hidden abilities. "For many parents," Sonne told me, "this was one of the few moments when they could be proud of their children." So he decided to ask potential employees to follow the assembly directions included in the Lego Mindstorms kits and watch them build the robots.

This turned out to be so revealing that assessing job skills in the autistic population has itself become part of Specialisterne's business, with local government sending about fifty people a year to the company for five-month evaluations. (Specialisterne considers some for consulting jobs; others might end up doing clerical work, mowing lawns, or performing other tasks for municipalities.) The Specialisterne evaluators place the candidates in groups for part of the time to see how well they work in teams, in addition to assessing the skills (reasoning, following directions, attending to details) that are naturally on display in a Mindstorms session. The assignments also reveal how a person handles trouble. More than once a candidate has become derailed because a Lego piece does not match the shade of gray depicted in the manual. Yet it is also not uncommon for a candidate to notice a struggling partner, stop, and patiently explain how to get back on track.

The Specialisterne school uses Legos, too. Frank Paulsen, a red-haired man with a thin beard who is the school's principal, told me about a session he once led in which he handed out small Lego boxes to a group of young men and asked them to build something that showed their lives. When the bricks had been snapped together, Paulsen asked each boy to say a few words. One boy didn't want to talk, saying his construction was "nothing." When Paulsen gathered his belongings to leave, however, the boy, his teacher by his side, seemed to want to stay. Paulsen tried to draw him out but failed. So Paulsen excused himself and stood up.

The boy grabbed Paulsen's arm. "Actually," he said, "I think I built my own life."

Paulsen eased back into his seat.

"This is me," the boy said, pointing to a skeleton penned in by a square structure with high walls. A gray chain hung from the back wall, and a drooping black net formed the roof. To the side, outside the wall, two figures—a man with a red baseball cap and a woman raising a clear goblet to her lips—stood by a translucent blue sphere filled with little gold coins. That, the boy continued, represented "normal life." In front of the skeleton were low walls between a pair of tan pillars, and a woman with a brown ponytail looked in, brandishing a yellow hairbrush. "That is my mom, and she is the only one who is allowed in the walls."

The boy's teacher was listening, astonished: In the years she'd known him, she told Paulsen later, she had never heard him discuss his inner life. Paulsen talked to the boy, now animated, for a quarter of an hour about the walls, and Paulsen suggested that perhaps the barriers could be removed. "I can't take down the walls," the boy concluded, "because there is so much danger outside of them."

In June, Sonne announced the opening of a U.S. headquarters in Wilmington, Delaware. The state's governor, Jack Markell, was there, as was a representative from CAI, the company that is Specialisterne's first real partner in the United States. The company says it plans to begin recruiting and training autistic software testers in Delaware next month, and if all goes well, it will expand the program to other states. Specialisterne is also talking with Microsoft about setting up a pilot program in Fargo, North Dakota, where it has a large software-development operation.

Tyler Cowen, an economist at George Mason University (and a regular contributor to the *New York Times*), published a much-discussed paper last year that addressed the ways that autistic workers are being drawn into the modern economy. The autistic worker, Cowen wrote, has an unusually wide variation in his or her skills, with higher highs and lower lows. Yet today, he argued, it is increasingly a worker's *greatest* skill, not his average skill level, that matters. As capitalism has grown more adept at disaggregating tasks, workers can focus on what they do best, and managers are challenged to make room for brilliant, if difficult, outliers. This march toward greater specialization, combined with the pressing need for expertise in science, technology, engineering, and mathematics, so-called STEM workers, suggests that the prospects for autistic workers will be on the rise in the coming decades. If the market can forgive people's weaknesses, then they will rise to the level of their natural gifts.

"Specialization is partly about making good use of the skills of people who have one type of skill in abundance but not necessarily others," says Daron Acemoglu, an economist at MIT and coauthor of *Why Nations Fail*. In other words, there is good money to be made doing the work that others do not have the skills for or are simply not interested in.

As Sonne tries to build up his business in the United States, though, he faces practical challenges. For one thing, in Denmark, the government helps cover some of the additional expense of managing autistic workers, and it pays Specialisterne so it can give its employees full-time salaries even though they work only part-time. Specialisterne pays its consultants in Denmark between $22 and $39 an hour, a rate negotiated with unions, and in Delaware it plans to start with salaries between $20 and $30 an hour. And while two Delaware charitable foundations have pledged $800,000 to Specialisterne, Sonne estimates that it will take $1.36 million, and three years, for the business to become self-sustaining.

Another challenge involves expectations. A new stereotype of autistic people as brainiacs endowed with quirky superminds is just as misguided as the old assumption that autistic people are mentally disabled, Sonne says. Autistic people, like everyone else, have diverse abilities and interests, and Specialisterne can't employ all of them. Most people Specialisterne evaluates in Denmark don't have the right qualities to be a consultant—they are too troubled, too reluctant to work in an office, or simply lack the particular skills Specialisterne requires. The company hires only about one in six of the men and women it assesses.

April Schnell, who is organizing a Specialisterne effort in the Midwest and has an autistic son, told me that she traveled to Copenhagen for a conference organized by the company for their volunteers from around the world. One day she and the others were given the Mindstorms challenges used to assess candidates.

As she struggled to solve one of the more difficult ones, she realized that her son, Tim, who is fifteen, would find the work uninteresting and probably too difficult: Specialisterne is not likely to be the answer for him. "I was just very aware, there is a gap here," she said. "My heart was a little sad."

One Friday evening, Sonne drove me to his house southwest of Copenhagen, navigating through whipping rain and the last clots of rush-hour traffic. Lars was waiting at the door to welcome us. Now sixteen, Lars evokes a Tolkien elf—thin and blond with exceptionally pale skin. He was outgoing from the start, eager to give me a tour of the house, yet he only glanced at my face.

Lars has the sweet demeanor of a much younger boy. Several times he affectionately rubbed his father's head, the hair a short thin fur, calling the bald spot "Mr. Moon." He gushed about trains, and at dinner Annette gently told him that we might not want to hear too much more about international conventions on track signals. I played Lars in a round of speed chess in the living room. There was never much doubt about the outcome, but at one point he issued an earnest warning: "Take care to not weaken your king's position unnecessarily." It was too late. After we put the pieces away, I complimented him on his final moves—an elegant and lethal attack with rooks, a bishop, and a knight—and he did a balletic twirl, arms out. I joked with his family about how crushed I felt in defeat, and Lars walked over and put a consoling hand on my shoulder. Perhaps, I suggested to Lars, I would be allowed a rematch? "No," he said simply.

When I asked Lars what he thought about his father's company, he said he has played with the Mindstorms robots but does not see himself working there. "I want to be a train driver," Lars announced. "It is the country's most beautiful job. You get to control a lot of horsepower. Who wouldn't want to do that?"

At the outset, it was Thorkil's aim to persuade Danish tech companies to hire his autistic employees. Now he wants all kinds of companies, all over the world, to learn from what Specialisterne is doing. He figures that if he is successful, then maybe a national railway will consider hiring a candidate as seemingly unlikely as his son, as long as he has the right skills.

Certainly he has seen how transformative getting the right job can be for the autistic workers themselves. Before coming to

Specialisterne, Iversen, who works at TDC, had not had a job for twelve years and spent the days sleeping and nights surfing the Internet. Niels Kjaer once worked as a physicist, receiving his diagnosis only after becoming clinically depressed when he didn't get an academic job. When he came to Specialisterne, where he works on improving technology that grades eggs as they pass by on a conveyor belt, he was on sick leave from a job driving a cab.

Christian Andersen, who works at Lundbeck, the pharmaceutical company, was bullied and beaten for years as a schoolboy. He received his diagnosis at age fifteen only because, fearing he might be suicidal, he checked himself into a hospital. After high school —inspired by a Hemingwayesque teacher who regaled his students with tales of outdoor exploits—Andersen tried a vocational school for landscaping. But he was overwhelmed by the requirement that he learn to drive. He tried another tech school but failed, became depressed, and had a breakdown in 2005. Andersen was living at home without prospects, playing video games. He couldn't even land a job at a grocery store. Later that year, his parents encouraged him to apply to Specialisterne.

I joined Andersen one morning on his commute to Lundbeck's headquarters across town. Riding on a yellow city bus, we talked about video games. He still loves Halo; Diablo 3 he finds frustrating. "You turn a corner and then—splat!—you are dead." As we drew closer to the office, our conversation drifted to his job. He spoke with surprising insight about the psychological importance of work. "I have grown very much as a person," Andersen told me. "I have become more confident and self-assured." The job allowed him to move out of his parents' house and into an apartment. After a while, Andersen informed me, he "started using body language." It's not something anyone taught him. He just watched people, he said, and "monkey see, monkey do."

When he started at Lundbeck, he was constantly anxious because he dreaded making an error. Now the stress grips him far less often and is readily dispelled with a phone call to a coach at Specialisterne. He admits to being proud, having come so far. He was touched to be invited recently to join his department for some after-work bowling. But he doesn't spend a lot of time thinking about these aspects of his employment anymore. "Of course it feels good," Andersen said, "but there is such a thing as 'here we go again.'" It's only a job, after all.

NATALIE ANGIER

The Life of Pi, and Other Infinities

FROM *The New York Times*

ON THIS DAY (December 31) that fetishizes finitude, that reminds us how rapidly our own earthly time share is shrinking, allow me to offer the modest comfort of infinities.

Yes, infinities, plural. The popular notion of infinity may be of a monolithic totality, the ultimate, unbounded big tent that goes on forever and subsumes everything in its path—time, the cosmos, your complete collection of old *Playbills*. Yet in the ever-evolving view of scientists, philosophers, and other scholars, there really is no single, implacable entity called infinity.

Instead, there are infinities, multiplicities of the limit-free that come in a vast variety of shapes, sizes, purposes, and charms. Some are tailored for mathematics, some for cosmology, others for theology; some are of such recent vintage their fontanels still feel soft. There are flat infinities, hunchback infinities, bubbling infinities, hyperboloid infinities. There are infinitely large sets of one kind of number, and even bigger, infinitely large sets of another kind of number.

There are the infinities of the everyday, as exemplified by the figure of pi, with its endless postdecimal tail of nonrepeating digits, and how about if we just round it off to 3.14159 and then serve pie on March 14 at 1:59 P.M.? Another stalwart of infinity shows up in the mathematics that gave us modernity: calculus.

"All the key concepts of calculus build on infinite processes of one form or another that take limits out to infinity," said Steven Strogatz, author of the recent book *The Joy of x: A Guided Tour of Math, from One to Infinity* and a professor of applied mathematics at Cornell. In calculus, he added, "infinity is your friend."

Yet worthy friends can come in prickly packages, and mathematicians have learned to handle infinity with care.

"Mathematicians find the concept of infinity so useful, but it can be quite subtle and quite dangerous," said Ian Stewart, a mathematics researcher at the University of Warwick in England and the author of *Visions of Infinity*, the latest of many books. "If you treat infinity like a normal number, you can come up with all sorts of nonsense, like saying infinity plus one is equal to infinity, and now we subtract infinity from each side and suddenly naught equals one. You can't be freewheeling in your use of infinity."

Then again, a very different sort of infinity may well be freewheeling you. Based on recent studies of the cosmic-microwave afterglow of the Big Bang, with which our known universe began 13.7 billion years ago, many cosmologists now believe that this observable universe is just a tiny, if relentlessly expanding, patch of spacetime embedded in a greater universal fabric that is, in a profound sense, infinite. It may be an infinitely large monoverse, or it may be an infinite bubble bath of infinitely budding and inflating multiverses, but infinite it is, and the implications of that infinity are appropriately huge.

"If you take a finite physical system and a finite set of states, and you have an infinite universe in which to sample them, to randomly explore all the possibilities, you will get duplicates," said Anthony Aguirre, an associate professor of physics who studies theoretical cosmology at the University of California, Santa Cruz.

Not just rough copies, either. "If the universe is big enough, you can go all the way," Aguirre said. "If I ask, will there be a planet like Earth with a person in Santa Cruz sitting at this colored desk, with every atom, every wave function exactly the same, if the universe is infinite the answer has to be yes."

In short, your doppelgängers may be out there and many variants, too, some with much better hair who can play Bach like Glenn Gould. A far less savory thought: there could be a configuration, Aguirre said, "where the Nazis won the war."

Given infinity's potential for troublemaking, it's small wonder the ancient Greeks abhorred the very notion of it.

"They viewed it with suspicion and hostility," said A. W. Moore, a professor of philosophy at Oxford University and the author of *The Infinite* (1990). The Greeks wildly favored tidy rational numbers that, by definition, can be defined as a ratio, or fraction—the way 0.75 equals ¾ and you're done with it—over patternless infinitums like the square root of 2.

On Pythagoras's Table of Opposites, "the finite" was listed along with masculinity and other good things in life, while "the infinite" topped the column of bad traits like femininity. "They saw it as a cosmic fight," Moore said, "with the finite constantly having to subjugate the infinite."

Aristotle helped put an end to the rampant infiniphobia by drawing a distinction between what he called "actual" infinity, something that would exist all at once at a given moment—which he declared an impossibility—and "potential" infinity, which would unfold over time and which he deemed perfectly intelligible. As a result, Moore said, "Aristotle believed in finite space and infinite time," and his ideas held sway for the next 2,000 years.

Newton and Leibniz began monkeying with notions of infinity when they invented calculus, which solves tricky problems of planetary motions and accelerating bodies by essentially breaking down curved orbits and changing velocities into infinite series of tiny straight lines and tiny uniform motions. "It turns out to be an incredibly powerful tool if you think of the world as being infinitely divisible," Strogatz said.

In the late nineteenth century, the great German mathematician Georg Cantor took on infinity not as a means to an end but as a subject worthy of rigorous study in itself. He demonstrated that there are many kinds of infinite sets, and some infinities are bigger than others. Hard as it may be to swallow, the set of all the possible decimal numbers between 1 and 2, being unlistable, turns out to be a bigger infinity than the set of all whole numbers from 1 to forever, which in principle can be listed.

In fact, many of Cantor's contemporaries didn't swallow, dismissing him as "a scientific charlatan," "laughable," and "wrong." Cantor died depressed and impoverished, but today his set theory is a flourishing branch of mathematics relevant to the study of large, chaotic systems like the weather, the economy, and human stupidity.

With his majestic theory of relativity, Einstein knitted together time and space, quashing old Aristotelian distinctions between actual and potential infinity and ushering in the contemporary era of infinity-seeking. Another advance came in the 1980s, when Alan

Guth introduced the idea of cosmic inflation, a kind of vacuum energy that vastly expanded the size of the universe soon after its fiery birth.

New theories suggest that such inflation may not have been a one-shot event but rather part of a runaway process called eternal inflation, an infinite ballooning and bubbling outward of this and possibly other universes.

Relativity and inflation theory, said Aguirre, "allow us to conceptualize things that would have seemed impossible before."

Time can be twisted, he said, "so from one point of view the universe is a finite thing that is growing into something infinite if you wait forever, but from another point of view it's always infinite."

Or maybe the universe is like Jorge Luis Borges's fastidiously imagined Library of Babel, composed of interminable numbers of hexagonal galleries with polished surfaces that "feign and promise infinity."

Or like the multiverse as envisioned in Tibetan Buddhism, "a vast system of 10^{59} universes, that together are called a Buddha Field," said Jonathan C. Gold, who studies Buddhist philosophy at Princeton.

The finite is nested within the infinite, and somewhere across the glittering, howling universal sample space of Buddha Field or Babel, your doppelgänger is hard at the keyboard, playing a Bach toccata.

ROBERT M. SAPOLSKY

Super Humanity

FROM *Scientific American*

SIT DOWN WITH an anthropologist to talk about the nature of humans, and you are likely to hear this chestnut: "Well, you have to remember that 99 percent of human history was spent on the open savanna in small hunter-gatherer bands." It's a classic cliché of science, and it's true. Indeed, those millions of ancestral years produced many of our hallmark traits—upright walking and big brains, for instance. Of course, those wildly useful evolutionary innovations came at a price: achy backs from our bipedal stance; existential despair from our large, self-contemplative cerebral cortex. As is so often the case with evolution, there is no free lunch.

Compounding the challenges of those trade-offs, the world we have invented—and quite recently in the grand scheme of things—is dramatically different from the one to which our bodies and minds are adapted. Have your dinner come to you (thanks to the pizza delivery guy) instead of chasing it down on foot; log in to Facebook to interact with your nearest and dearest instead of spending the better part of every day with them for your whole life. But this is where the utility of the anthropologist's cliché for explaining the human condition ends.

The reason for this mismatch between the setting we evolved to live in and the situations we encounter in our modern era derives from another defining characteristic of our kind, arguably the most important one: our impulse to push beyond the limitations evolution imposed on us by developing tools to make us faster, smarter, longer-lived. Science is one such tool—an invention that requires us to break out of our Stone Age seeing-is-believing mind-set so that we can clear the next hurdle we encounter, be it a pandemic flu or climate change. You could call it the ultimate expression of humanity's singular drive to aspire to be better than we are.

Human Oddities

To understand how natural selection molded us into the unique primates we have become, let us return to the ancestral savanna. That open terrain differed considerably from the woodlands our ape forebears called home. For one thing, the savanna sun blazed hotter; for another, nutritious plant foods were scarcer. In response, our predecessors lost their thick body hair to keep cool. And their molars dwindled as they abandoned a tough vegetarian diet for one focused in part on meat from grassland grazers—so much so that our molars are now nearly useless, with barely any grinding surface.

Meanwhile the selective demands of food scarcities sculpted our distant forebears into having a body that was extremely thrifty and good at storing calories. Now, having inherited that same metabolism, we hunt and gather Big Macs as diabetes becomes a worldwide scourge. Or consider how our immune systems evolved in a world where one hardly ever encountered someone carrying a novel pathogen. Today, if you sneeze near someone in an airport, your rhinovirus could be set free twelve time zones away by the next day.

Our human oddities abound where behavior is concerned. By primate standards, we are neither fish nor fowl in lots of ways. One example is particularly interesting. Primate species generally fall into two distinct types: on one hand, there are pair-bonding species, in which females and males form stable, long-lasting pairs that practice social and sexual monogamy. Monogamous males do some or even most of the caring for the young, and females and males in these species are roughly the same size and look very similar. Gibbons and numerous South American monkeys show this pattern. "Tournament" species take the opposite tack: females do all the child care, whereas males are far larger and come with all kinds of flashy displays of peacockery—namely, gaudy, conspicuous facial coloration and silver backs. These tournament males spend a ridiculous percentage of their time enmeshed in aggressive posturing. And then there are humans, who, by every anatomical, physiological, and even genetic measure, are neither classic pair-bonding nor tournament creatures and instead lie stuck and confused somewhere in the middle.

Yet in another behavioral regard, humans are textbook primates: we are intensely social, and our fanciest types of intelligence are the social kinds. We primates may have circumstances where a complex mathematical instance of transitivity bewilders us, but it is simple for us to figure out that if person A dominates B, and B dominates C, then C had better grovel and submissively stick his butt up in the air when A shows up. We can follow extraordinarily complex scenarios of social interaction and figure out if a social contract has been violated (and are better at detecting someone cheating than someone being overly generous). And we are peerless when it comes to facial recognition: we even have an area of the cortex in the fusiform gyrus that specializes in this activity.

The selective advantages of evolving a highly social brain are obvious. It paved the way for us to fine-tune our capacities for reading one another's mental states, to excel at social manipulation, and to adeptly deceive and attract potential mates and supporters. Among Americans, the extent of social intelligence in youth is a better predictor of our adult success in the occupational world than are SAT scores.

Indeed, when it comes to social intelligence in primates, humans reign supreme. The social-brain hypothesis of primate evolution is built on the fact that across primate species, the percentage of the brain devoted to the neocortex correlates with the average size of the social group of that species. This correlation is more dramatic in humans (using the group sizes found in traditional societies) than in any other primate species. In other words, the most distinctively primate part of the human brain coevolved with the demands of keeping track of who is not getting along with whom, who is tanking in the dominance hierarchy, and what couple is furtively messing around when they should not be.

Like our bodies, our brains and behaviors, sculpted in our distant hunter-gatherer past, must also accommodate a very different present. We can live thousands of miles away from where we were born. We can kill someone without ever seeing his face. We encounter more people standing on line for Space Mountain at Disneyland than our ancestors encountered in a lifetime. My God, we can even look at a picture of someone and feel lust despite not knowing what that person smells like—how weird is that for a mammal?

Beyond Limits

The fact that we have created and are thriving in this unrecognizable world proves a point—namely, that it is in our nature to be unconstrained by our nature. We are no strangers to going out of bounds. Science is one of the strangest, newest domains where we challenge our hominid limits. Some of the most dramatic ways in which our world has been transformed are the direct products of science, and the challenges there are obvious. Just consider those proto-geneticists who managed to domesticate some plants and animals—an invention that brought revolutionary gains in food but that now threatens to strip the planet of its natural resources.

On a more abstract plane, science tests our sense of what is the norm, what counts as better than well. It challenges our sense of who we are. Thanks to science, human life expectancy keeps extending, our average height increases, our scores on standardized tests of intelligence improve. Thanks to science, every world record for a sporting event is eventually surpassed.

As science pushes the boundaries in these domains, what is surprising is how little these changes have changed us. No matter how long we can expect to live, we still must die, there will still be a leading cause of death, and we will still feel that our loved ones were taken from us too soon. And when it comes to humans becoming, on average, smarter, taller, and better at athletics, there is a problem: Who cares about the average? As individuals, we want to individually be better than other individuals. Our brain is invidious, comparative, more interested in contrasts than absolutes. That state begins with sensory systems that often do not tell us about the quality of a stimulus but instead about the quality relative to the stimuli that surround it. For example, the retina contains cells that do not so much respond to a color as to a color in the context of its proximity to its "opposite" color (red versus green, for instance). Although we may all want to be smart, we mostly want to be smarter than our neighbor. The same is true for athletes, which raises a question that has long been pertinent to hominids: How fast do you have to run to evade a lion? And the answer always is: faster than the person next to you.

Still, science most asks us to push our limits when it comes to the kinds of questions we ask. I see four particular types. The first has to do with the frequently asocial nature of science. By this I am not referring to the solitary task of some types of scientific inquiry, the scientist slaving away alone at three in the morning. I mean that science often asks us to be really interested in inanimate things. There are obviously plenty of exceptions to this rule—primatologists sit around and gossip at night about the foibles and peccadilloes of their monkeys; the paleontologist Louis Leakey used to refer to his favorite fossil skull as "Dear Boy." Yet some realms of science consider extremely inanimate issues—astrophysicists trying to discover planets in other solar systems, for instance. Science often requires our social, hominid brain to be passionate about some pretty unlikely subjects.

Science pushes our envelope in a second way when we contemplate the likes of quantum mechanics, nanotechnology, and particle physics, which ask us to believe in things that we cannot see. I spent my graduate-school years pipetting fluids from one test tube to another, measuring levels of things like hormones and neurotransmitters. If I had stopped and thought about it, it would have seemed completely implausible that there actually are such things as hormones and neurotransmitters. That implausibility is the reason why so many of us lab scientists who measure or clone or inject invisible things get the most excited when we get to play with dry ice.

Science, by the nature of the questions it can generate, can push the bounds of our hominid credulity in a third way. We are unmatched in the animal kingdom when it comes to remembering the distant past, when it comes to having a sense of the future. These skills have limits, however. Traditionally our hunter-gatherer forebears may have remembered something their grandmother was told by her grandmother, or they may have imagined the course of a generation or two that would outlive them. But science sometimes asks us to ponder processes that emerge over time spans without precedent. When will the next ice age come? Will Gondwana ever reunite? Will cockroaches rule us in a million years?

Everything about our hominid minds argues against the idea that there are processes that take that long or that such processes could be interesting. We and other primates are creatures of steep temporal discounting—getting ten dollars or ten pellets of monkey chow right now is more appealing than waiting until tomorrow for eleven, and the dopamine-reward pathways in our brain light up on brain-imaging tests when we go for the impulsive immediate reward. It seems most of us would rather have half a piece of stale popcorn next week than wait 1,000 years to win a bet about a key hypothesis in plate tectonics.

Then there are scientific questions that stretch our limits in the most profound ways. These are quandaries of dazzling abstractness: Does free will exist? How does consciousness work? Are there things that are impossible to know?

It is tempting to fall for an easy insight here, which is that our Paleolithic minds give up on challenges like these and just turf them to the gods to contemplate. The problem is the human propensity toward creating gods in our own image (one fascinating example being that autistic individuals who are religious often have an image of an asocial god, one who is primarily concerned with the likes of keeping atoms from flying apart). Throughout the history of humans inventing deities, few of these gods had a gargantuan capacity for the abstract. Instead they had familiar appetites. No traditional deities would be particularly interested in chewing the fat with Gödel about knowingness or rolling dice with Einstein (or not rolling the dice, as it were). They would be much more into having the biggest ox sacrificed to them and scoring with the most forest nymphs.

The very scientific process defies our basic hominid limits. It asks us to care intensely about tiny, even invisible, things, things that do not breathe or move, things vast distances away from us in space and time. It encourages us to care about subjects that would bore the crap out of Thor or Baal. It is one of the most challenging things that we have come up with. No wonder all those nerd-detector alarms would go off back in middle school when we were spotted reading a magazine like *Scientific American*. This venture of doing, thinking, caring about science is not for the faint-hearted —we are far better adapted to face saber-toothed cats—and yet here we are, reinventing the world and striving to improve our lot in life one scientific question at a time. It's our human nature.

KATHERINE HARMON

The Patient Scientist

FROM *Scientific American*

PEERING THROUGH A microscope at a plate of cells one day, Ralph M. Steinman spied something no one had ever seen before. It was the early 1970s, and he was a researcher at Rockefeller University on Manhattan's Upper East Side. At the time, scientists were still piecing together the basic building blocks of the immune system. They had figured out that there are B cells, white blood cells that help to identify foreign invaders, and T cells, another type of white blood cell that attacks those invaders. What puzzled them, however, was what triggered those T cells and B cells to go to work in the first place. Steinman glimpsed what he thought might be the missing piece: strange, spindly-armed cells unlike any he had ever noticed.

His intuition turned out to be correct. These dendritic cells, as Steinman named them, are now thought to play a crucial role in detecting invaders in the body and initiating an immune response against them. They snag interlopers with their arms, ingest them, and carry them back to other types of immune cells—in effect, "teaching" them what to attack. It was a landmark discovery that explained in unprecedented detail how vaccines worked, and it propelled Steinman into the top tiers of his profession.

In many ways, Steinman's story is typical: brilliant scientist makes major discovery that inspires a new generation of researchers. Indeed, his insight was remarkable for its implications, both for science and for him personally.

Over the years Steinman came to believe that dendritic cells were a crucial weapon for tackling some of the most loathed diseases, from cancer to HIV. He and his global network of colleagues seemed to be well on the way to proving him correct when Steinman's story took an unusual turn.

In 2007 he was diagnosed with pancreatic cancer, an unforgiving disease that kills four out of five patients within a year. In the end, the cells he discovered at the start of his career, and the friends he made along the way, would not only help him fight his cancer but would extend his life just long enough for him to earn the Nobel Prize. He died this past September, three days before a flashing light on his cell phone alerted his family that he had won.

A Prepared Mind

Steinman did not encounter serious biology until he arrived as a student at McGill University. As soon as he did, though, he was hooked, and it was his fascination with the minuscule world of the immune cell that would bring him to the lab of Zanvil A. Cohn at Rockefeller. In his office Steinman would later display a quote from the famous nineteenth-century microbiologist and vaccinologist Louis Pasteur: *Le hazard ne favorise que les esprits préparés,* which is often translated as "Chance favors the prepared mind." Says Sarah Schlesinger, a longtime colleague and friend of Steinman's, "Ralph was exceedingly well prepared, so he was poised to make a discovery. But with that said, he intuited that these were important," she says of the cells. It was that intuition and a confidence in observation that enabled him to make his seminal discovery—and eventually win the admiration of colleagues.

After he first spotted dendritic cells, Steinman spent the next two decades convincing the scientific community of their significance, defining how they worked and how researchers could work with them. "He *fought*—there's really no other word for it—to convince people that they were a distinct entity," says Schlesinger, who came to work at Steinman's lab in 1977, when she was still in high school. Even then, she says, people in the same lab were not convinced that these dendritic cells existed because they were difficult to enrich into larger batches. At the time, Steinman was still working at the bench, and Schlesinger recalls sitting with him at a two-headed microscope, examining the cells. "He just loved to look at them," she says, smiling at the memory. "There was such a joy in all of the little discoveries that he made."

By the 1980s Steinman, who had trained as a physician, started to look for ways his dendritic cell discovery could be applied more directly to help people. Over the next few decades, as the cells became more widely accepted, his lab expanded its focus to include research into dendritic cell–based vaccines for HIV and tuberculosis, as well as research into cancer treatment. For illnesses such as influenza or smallpox that could already be prevented with vaccines, those who survive natural exposure may develop a lifelong immunity. HIV, TB, and cancer presented a greater challenge because they seemed to be better at overcoming the immune system — even, in the case of HIV, hijacking dendritic cells to do its dirty work. "Ralph would say, 'We have to be smarter than nature,'" Schlesinger says. That meant helping the dendritic cells by giving them more targeted information about the virus or tumor against which the immune system needed to form an attack.

In the 1990s, working with Madhav Dhodapkar, now at Yale University, and Nina Bhardwaj, now at New York University, Steinman created a process for extracting dendritic cells from the blood and priming them with antigens — telltale protein fragments — from infections, such as influenza and tetanus, and then placing them back in the body to create a stronger immunity. This technology served as the basis for a prostate cancer vaccine called Provenge that was approved in 2010 and has been shown to extend the life of terminally ill patients — if only by a few months.

The Final Experiment

In early 2007 Steinman was away in Colorado at a scientific meeting, a trip that he had turned into a family ski vacation, when he and his twin daughters all had what seemed like a stomach bug. His daughters recovered quickly, but his illness lingered. Soon after he returned home, he developed jaundice. In the third week of March he went in for a CT scan, and radiologists found a tumor in his pancreas. By then, it had already spread to his lymph nodes. He knew his odds of survival were slim: about 80 percent of pancreatic cancer patients die within a year.

"When he first told us, he said, 'Do not Google this — just listen to me,'" his daughter Alexis recalls. She felt like someone had punched her. "He really expressed to the family that while it was a very drastic disease, he was in a very good position," she says. Unlike the average cancer patient, Steinman had access to many of the top immunologists and oncologists on the planet—and, perhaps even more important, to their most promising therapies.

When Schlesinger heard the news, she was devastated. And she quickly rallied to her mentor's side. She, Steinman, and their close Rockefeller colleague Michel Nussenzweig began making phone calls, sharing the news with colleagues across the globe. Steinman was convinced that the surest way to be cured of any tumor was to develop immunity against it through his own dendritic cells. They had a limited amount of time to prove him right.

One of the early calls Steinman made after his diagnosis was to his longtime collaborator Jacques Banchereau, who now directs the Baylor Institute for Immunology Research in Dallas. Banchereau then picked up the phone to call Baylor researcher Anna Karolina Palucka, who had known Steinman since the 1990s. Although she had an experimental vaccine in the works that she thought could help Steinman, she struggled with the personal challenge of trying "to compartmentalize the friend, the patient, and the scientist."

For her part, Schlesinger called Charles Nicolette, a friend and collaborator of many years and the chief scientific officer of Argos Therapeutics, an RNA-based drug company in Durham, North Carolina, that Steinman had cofounded. Nicolette, reeling from the news, mobilized his own colleagues within minutes of hanging up the phone.

Nicolette's group had developed a dendritic cell vaccine that was in a phase II (intermediate-stage) clinical trial to treat advanced kidney cancer. Argos's therapy endeavors to enlist a patient's own dendritic cells against a cancer by exposing them to genetic material from the tumor, which induces them to rally T cells to mount a proper attack.

Steinman was scheduled to have part of his pancreas removed during the first week of April 2007—a surgery known as a Whipple procedure, which is part of a more traditional treatment for his prognosis. Nicolette would need part of that tumor to draw up his vaccine, which left him just days to get the U.S. Food and Drug Administration to approve Steinman's entry into his trial, permission the team was able to secure just in time.

With the tumor cells secured and while the Argos treatment was brewing, a process that would take months, Steinman started in on other therapies. Soon after his surgery, he went on standard Gemcitibine-based chemotherapy, and then, in the late summer, he enrolled in a trial of GVAX, a dendritic cell–based vaccine that was being tested to treat pancreatic cancer. Codeveloped by Elizabeth Jaffee of Johns Hopkins University and administered at the Dana-Farber/Harvard Cancer Center, the vaccine uses a generic tumor antigen, as the Provenge prostate cancer vaccine does. In an earlier phase II trial, pancreatic cancer patients who had received the vaccine lived an average of four months longer than those who had not, and some ended up living for years. So for two months, starting in the late summer, Schlesinger traveled with Steinman to Boston almost every week. "I remember walking in Boston on a day like this," she says, looking out of her corner office window into the clear, paling blue October afternoon sky, "thinking, He's not going to see another fall, and I was so sad."

But fall came and went, and Steinman remained in relatively good health. In September 2007 he received the Albert Lasker Award for Basic Medical Research, considered by many to be a precursor to the Nobel, and he sat for a series of video interviews. In them he elaborated on the promise of dendritic cells to fight cancer, noting that an immune attack is highly directed, highly specific, and, unlike chemotherapy, nontoxic. "I think this provides the potential for a whole new type of therapy in cancer," he said. "But we need research and patience to discover the rules, to discover the principles."

At times Steinman showed more patience than his colleagues would have liked. He had initially argued for a very slow course of treatment for himself so that his team could monitor his immune response after each therapy before beginning the next. But Schlesinger and Nussenzweig eventually convinced him that they simply did not have the time. If he died, the experiment and data collection were over.

By November 2007 the Argos vaccine, made by infusing cells taken from Steinman's blood with genetic material extracted from his tumor, was ready and waiting. Steinman had just finished with a chemo treatment, and he enrolled in Argos's renal-cell carcinoma trial under a single-patient study protocol.

In early 2008 Steinman followed up with Palucka's vaccine, which was being developed for melanoma. It incorporated a selection of tumor-specific peptides (protein fragments), so she suspected it could be repurposed to target Steinman's cancer by using peptides from his tumor in place of antigens from melanoma.

Other offers for experimental treatments poured in from all over the world. "Everybody who could brought the best they could," Palucka says. Steinman's decades of collegial work had united the field, and now that network of scientists turned to help one of their own. "People think of science as a solitary process. In fact, it's an extremely social process," Schlesinger says. The "social nature of our work facilitated the forthcoming of these tremendous intellectual resources."

In addition to standard treatment, Steinman ended up enrolled —under a special patient provision—in four ongoing clinical trials of various dendritic cell–based cancer treatments, most of which were not even being tested for pancreatic cancer, along with several other experimental immunotherapy and chemo treatments. Schlesinger, a member of the Rockefeller Institutional Review Board (IRB), steered his treatment through all the necessary IRB and FDA channels, making sure the standard protocols were followed. She also personally gave Steinman his vaccines whenever they could be administered at Rockefeller.

Steinman ran his own grand experiment the way he ran others in the lab—always carefully collecting data, evaluating the evidence, and doling out instructions. Schlesinger still has e-mail chains from the period, Steinman's messages coming back in all capital letters per his style. He kept particularly close tabs on how his own body was responding to treatment. In 2008, during his time on Palucka's therapy, she came for a visit to New York City. After Schlesinger had given Steinman his dose of the vaccine, the three of them went out to dinner. On finishing their meal, Steinman insisted they stop by Palucka's hotel so that he could show them the welt developing on his leg around the injection site. "He was so enthused about it," Schlesinger says. "He said, 'Those are T cells'"—indicating that his body was having an immune response to the vaccine—"'that's great!'"

The local swelling showed that Steinman's body was reacting to the vaccine, although, Palucka says, she cannot be certain it was tumor-specific T cells that had been mobilized. As she points out, all vaccines work through dendritic cells, but the difference with her therapy and the others that Steinman tried was that rather than leaving exposure up to chance, researchers manipulated the dendritic cells outside of the body to improve the odds that they would train T cells to attack the tumor. When Schlesinger was not on hand to see the evidence for herself, she says, "he would send me these descriptions of the vaccination sites with great enthusiasm," including information about the appearance and size of the sites—and even how each one felt.

His tumor marker, the level of a protein that indicates the progress of a cancer (which fluctuated throughout the course of his treatment), became a barometer for his attitude. The second time the marker went down, he sent an e-mail out with the subject line "We've repeated the experiment," the glee of which was apparent to those who knew his joy in a scientific triumph.

But the good news that satisfied Steinman the patient was never good enough to satisfy Steinman the scientist. The knowledge that his one-person experiment was hardly a scientific one frustrated him no end. With the experimental treatments administered so close to one another—and interspersed with traditional chemotherapy—it was impossible to know what sent his tumor biomarker downward.

Nevertheless, Steinman generated some interesting data points along the way. During one of Palucka's immune-monitoring tests during his treatment, she found that some 8 percent of cells known as CD8 T cells (also called killer T cells) were specifically targeted to his tumor. That might not sound like a lot, but given all the potential pathogens that the body can encounter and mount an attack against, 8 percent "is a huge number," Schlesinger says. "So something immunized him—or some combination of things immunized him."

A Death, Days Too Soon

Steinman and his wife, Claudia, traveled to Italy to celebrate their fortieth wedding anniversary in June 2011—just two months after what he referred to as his fourth "Whipple-versary," in honor of his April 2007 surgery. Already he had far surpassed the average survival of a person with his type of cancer.

In mid-September 2011, Steinman was still working at the lab, and arrangements had been made for him to restart the Argos treatment. Then Steinman fell ill with pneumonia. "When he was admitted to the hospital, he said, 'I might not make it out of here,'" Alexis recalls. But after her father's four and a half years of good health, she found it hard to believe she would have only days left with him. He was still reviewing data from Rockefeller as late as September 24. On Friday, September 30, he died at the age of sixty-eight from respiratory failure caused by pneumonia, which his cancer-weakened body could no longer fend off.

His family struggled with how to even begin to tell his vast network of friends and colleagues around the globe. They planned to visit his old lab—where he had been working until so recently—to tell those there on Monday, October 3. But early that morning, before any of them were awake, Stockholm called. Steinman's BlackBerry, on silent, was with his wife. In a fitful early-morning sleep, she glanced over to see a new-message light blinking. Just then an e-mail popped up, politely informing Steinman that he had won the 2011 Nobel Prize in Physiology or Medicine.

The first response was that "we all collectively screamed the 'f' word," Alexis says. Her next thought was "Let's go wake up Dad."

But for the rest of the world, nothing about the Nobel committee's announcement seemed amiss—articles were written, statements were issued about Steinman and the two other recipients, Bruce Beutler of the Scripps Research Institute and Jules Hoffmann of the French National Center for Scientific Research—until a few hours later, when news of Steinman's death surfaced. The prize rules state that it cannot be given posthumously, but if a laureate dies between the October announcement and the award ceremony in December, he or she can remain on the list. This odd timing threw the committee into a closely followed deliberation before it announced, late in the day, that he would remain a prize recipient.

Just days after Steinman's Nobel was announced and news of his death hit the media, pancreatic cancer also claimed the life of Apple's cofounder and CEO, Steve Jobs. Jobs, ill with a rare, slower-growing form of the disease—a neuroendocrine tumor—lived for eight years after his diagnosis, more of an average survival time for a patient with his form of the disease. Steinman's survival, though, far surpassed what was expected. "There's no question something extended his life," Schlesinger says.

Now researchers are working to figure out what it was. In early 2012 Baylor will be dedicating the Ralph Steinman Center for

Cancer Vaccines, and Palucka is developing a clinical trial to treat pancreatic cancer patients with the same vaccine that she helped create for Steinman. At Argos, Nicolette is pursuing their kidney cancer vaccine full steam ahead: "There's a sense of duty to Ralph to see this through." This month they plan to launch a phase III clinical trial of the renal cancer vaccine Steinman tried.

For her part, Schlesinger believes her colleagues' interventions made a contribution in the end. "The scientific message is: immunity makes a difference," she says. But the final lesson is one Steinman liked to preach. "He used to tell people, 'There are so many other things left to discover,'" she recalls. "And there are."

NATHANIEL RICH

Can a Jellyfish Unlock the Secret of Immortality?

FROM *The New York Times Magazine*

AFTER MORE THAN 4,000 years—almost since the dawn of recorded time, when Utnapishtim told Gilgamesh that the secret to immortality lay in a coral found on the ocean floor—man finally discovered eternal life in 1988. He found it, in fact, on the ocean floor. The discovery was made unwittingly by Christian Sommer, a German marine-biology student in his early twenties. He was spending the summer in Rapallo, a small city on the Italian Riviera, where exactly one century earlier Friedrich Nietzsche conceived *Thus Spoke Zarathustra:* "Everything goes, everything comes back; eternally rolls the wheel of being. Everything dies, everything blossoms again . . ."

Sommer was conducting research on hydrozoans, small invertebrates that, depending on their stage in the life cycle, resemble either a jellyfish or a soft coral. Every morning, Sommer went snorkeling in the turquoise water off the cliffs of Portofino. He scanned the ocean floor for hydrozoans, gathering them with plankton nets. Among the hundreds of organisms he collected was a tiny, relatively obscure species known to biologists as *Turritopsis dohrnii*. Today it is more commonly known as the immortal jellyfish.

Sommer kept his hydrozoans in petri dishes and observed their reproduction habits. After several days he noticed that his *Turritopsis dohrnii* was behaving in a very peculiar manner, for which he could hypothesize no earthly explanation. Plainly speaking, it refused to die. It appeared to age in reverse, growing younger and younger until it reached its earliest stage of development, at which point it began its life cycle anew.

Sommer was baffled by this development but didn't immediately grasp its significance. (It was nearly a decade before the word "immortal" was first used to describe the species.) But several biologists in Genoa, fascinated by Sommer's finding, continued to study the species, and in 1996 they published a paper called "Reversing the Life Cycle." The scientists described how the species —at any stage of its development—could transform itself back to a polyp, the organism's earliest stage of life, "thus escaping death and achieving potential immortality." This finding appeared to debunk the most fundamental law of the natural world—you are born, and then you die.

One of the paper's authors, Ferdinando Boero, likened the *Turritopsis* to a butterfly that, instead of dying, turns back into a caterpillar. Another metaphor is a chicken that transforms into an egg, which gives birth to another chicken. The anthropomorphic analogy is that of an old man who grows younger and younger until he is again a fetus. For this reason *Turritopsis dohrnii* is often referred to as the Benjamin Button jellyfish.

Yet the publication of "Reversing the Life Cycle" barely registered outside the academic world. You might expect that, having learned of the existence of immortal life, man would dedicate colossal resources to learning how the immortal jellyfish performs its trick. You might expect that biotech multinationals would vie to copyright its genome; that a vast coalition of research scientists would seek to determine the mechanisms by which its cells aged in reverse; that pharmaceutical firms would try to appropriate its lessons for the purposes of human medicine; that governments would broker international accords to govern the future use of rejuvenating technology. But none of this happened.

Some progress has been made, however, in the quarter century since Christian Sommer's discovery. We now know, for instance, that the rejuvenation of *Turritopsis dohrnii* and some other members of the genus is caused by environmental stress or physical assault. We know that during rejuvenation it undergoes cellular transdifferentiation, an unusual process by which one type of cell is converted into another—a skin cell into a nerve cell, for instance. (The same process occurs in human stem cells.) We also know that, in recent decades, the immortal jellyfish has rapidly spread throughout the world's oceans in what Maria Pia Miglietta, a biology professor at Notre Dame, calls "a silent invasion." The jellyfish has been "hitchhiking" on cargo ships that use seawater for ballast. *Turritopsis* has now been observed not only in the Mediterranean but also off the coasts of Panama, Spain, Florida, and Japan. The jellyfish seems able to survive, and proliferate, in every ocean in the world. It is possible to imagine a distant future in which most other species of life are extinct but the ocean consists overwhelmingly of immortal jellyfish, a great gelatin consciousness everlasting.

But we still don't understand how it ages in reverse. There are several reasons for our ignorance, all of them maddeningly unsatisfying. There are, to begin with, very few specialists in the world committed to conducting the necessary experiments. "Finding really good hydroid experts is very difficult," says James Carlton, a professor of marine sciences at Williams College and the director of the Williams-Mystic Maritime Studies Program. "You're lucky to have one or two people in a country." He cited this as an example of a phenomenon he calls the Small's Rule: small-bodied organisms are poorly studied relative to larger-bodied organisms. There are significantly more crab experts, for instance, than hydroid experts.

But the most frustrating explanation for our dearth of knowledge about the immortal jellyfish is of a more technical nature. The genus, it turns out, is extraordinarily difficult to culture in a laboratory. It requires close attention and an enormous amount of repetitive, tedious labor; even then, it is under only certain favorable conditions, most of which are still unknown to biologists, that a *Turritopsis* will produce offspring.

In fact there is just one scientist who has been culturing *Turritopsis* polyps in his lab consistently. He works alone, without major financing or a staff, in a cramped office in Shirahama, a sleepy beach town in Wakayama Prefecture, Japan, four hours south of Kyoto. The scientist's name is Shin Kubota, and he is, for the time being, our best chance for understanding this unique strand of biological immortality.

Many marine biologists are reluctant to make such grand claims about *Turritopsis*'s promise for human medicine. "That's a question for journalists," Boero said (to a journalist) in 2009. "I prefer to focus on a slightly more rational form of science."

Kubota, however, has no such compunction. "*Turritopsis* application for human beings is the most wonderful dream of mankind," he told me the first time I called him. "Once we determine how the jellyfish rejuvenates itself, we should achieve very great things. My opinion is that we will evolve and become immortal ourselves."

I decided I'd better book a ticket to Japan.

One of Shirahama's main attractions is its crescent-shaped white-sand beach; the name Shirahama means "white beach." But in recent decades, the beach has been disappearing. In the 1960s, when Shirahama was connected by rail to Osaka, the city became a popular tourist destination, and blocky white hotel towers were erected along the coastal road. The increased development accelerated erosion, and the famous sand began to wash into the sea. Worried that the town of White Beach would lose its white beach, according to a city official, Wakayama Prefecture began in 1989 to import sand from Perth, Australia, 4,700 miles away. Over fifteen years, Shirahama dumped 745,000 cubic meters of Aussie sand on its beach, preserving its eternal whiteness—at least for now.

Shirahama is full of timeless natural wonders that are failing the test of time. Visible just off the coast is Engetsu island, a sublime arched sandstone formation that looks like a doughnut dunked halfway into a glass of milk. At dusk, tourists gather at a point on the coastal road where, on certain days, the arch perfectly frames the setting sun. Arches are temporary geological phenomena; they are created by erosion, and erosion ultimately causes them to collapse. Fearing the loss of Engetsu, the local government is trying to restrain it from deteriorating any further by reinforcing the arch with a harness of mortar and grout. A large scaffold now extends beneath the arch, and from the shore construction workers can be seen, tiny flyspecks against the sparkling sea, paving the rock.

Engetsu is nearly matched in beauty by Sandanbeki, a series of striated cliffs farther down the coast that drop 165 feet into turbulent surf. Beneath Sandanbeki lies a cavern that local pirates used as a secret lair more than a thousand years ago. Today the cliffs are one of the world's most famous suicide spots. A sign on the edge serves as a warning to those contemplating their own mortality: "Wait a minute. A dead flower will never bloom."

But Shirahama is best known for its *onsen,* saltwater hot springs that are believed to increase longevity. There are larger, well-appointed ones inside resort hotels, smaller tubs that are free to the public, and ancient bathhouses in cramped huts along the curving coastal road. You can tell from a block away that you are approaching an onsen because you can smell the sulfur.

Each morning, Shin Kubota, who is sixty, visits Muronoyu, a simple onsen popular with the city's oldest citizens that traces its history back 1,350 years. "Onsen activates your metabolism and cleans away the dead skin," Kubota says. "It strongly contributes to longevity." At 8:30 A.M., he drives fifteen minutes up the coast, past the white beach, where the land narrows to a promontory that extends like a pointing, arthritic finger, separating Kanayama Bay from the larger Tanabe Bay. At the end of this promontory stands Kyoto University's Seto Marine Biological Laboratory, a damp, two-story concrete block. Though it has several classrooms, dozens of offices, and long hallways, the building often has the appearance of being completely empty. The few scientists on staff spend much of their time diving in the bay, collecting samples. Kubota, however, visits his office every single day. He must, or his immortal jellyfish will starve.

The world's only captive population of immortal jellyfish lives in petri dishes arrayed haphazardly on several shelves of a small refrigerator in Kubota's office. Like most hydrozoans, *Turritopsis* passes through two main stages of life, polyp and medusa. A polyp resembles a sprig of dill, with spindly stalks that branch and fork and terminate in buds. When these buds swell, they sprout not flowers but medusas. A medusa has a bell-shaped dome and dangling tentacles. Any layperson would identify it as a jellyfish, though it is not the kind you see at the beach. Those belong to a different taxonomic group, Scyphozoa, and tend to spend most of their lives as jellyfish; hydrozoans have briefer medusa phases. An adult medusa produces eggs or sperm, which combine to create larvae that form new polyps. In other hydroid species, the medusa dies after it spawns. A *Turritopsis* medusa, however, sinks to the bottom of the ocean floor, where its body folds in on itself—assuming the jellyfish equivalent of the fetal position. The bell reabsorbs the tentacles, and then it degenerates further until it becomes a gelatinous blob. Over the course of several days, this blob forms an outer shell. Next it shoots out stolons, which resemble roots. The stolons lengthen and become a polyp. The new polyp produces new medusas, and the process begins again.

Kubota estimates that his menagerie contains at least 100 specimens, about three to a petri dish. "They are very tiny," Kubota, the proud papa, said. "Very cute." It *is* cute, the immortal jellyfish. An adult medusa is about the size of a trimmed pinkie fingernail. It trails scores of hairlike tentacles. Medusas found in cooler waters have a bright scarlet bell, but more commonly the medusa is translucent white, its contours so fine that under a microscope it looks like a line drawing. It spends most of its time floating languidly in the water. It's in no rush.

For the last fifteen years, Kubota has spent at least three hours a day caring for his brood. Having observed him over the course of a week, I can confirm that it is grueling, tedious work. When he arrives at his office, he removes each petri dish from the refrigerator, one at a time, and changes the water. Then he examines his specimens under a microscope. He wants to make sure that the medusas look healthy: that they are swimming gracefully; that their bells are unclouded; and that they are digesting their food. He feeds them artemia cysts—dried brine shrimp eggs harvested from the Great Salt Lake in Utah. Though the cysts are tiny, barely visible to the naked eye, they are often too large for a medusa to digest. In these cases Kubota, squinting through the microscope, must slice the egg into pieces with two fine-point needles, the way a father might slice his toddler's hamburger into bite-size chunks. The work causes Kubota to growl and cluck his tongue.

"Eat by yourself!" he yells at one medusa. "You are not a baby!" Then he laughs heartily. It's an infectious, ratcheting laugh that makes his round face even rounder, the wrinkles describing circles around his eyes and mouth.

It is a full-time job, caring for the immortal jellyfish. When traveling abroad for academic conferences, Kubota has had to carry the medusas with him in a portable cooler. (In recent years he has been invited to deliver lectures in Cape Town; Xiamen, China; Lawrence, Kansas; and Plymouth, England.) He also travels to Kyoto when he is obligated to attend administrative meetings at the university, but he returns the same night, an eight-hour round trip, in order not to miss a feeding.

Turritopsis is not the only focus of his research. He is a prolific author of scientific papers and articles, having published fifty-two in 2011 alone, many based on observations he makes on a private beach fronting the Seto Lab and in a small harbor on the coastal road. Every afternoon, after Kubota has finished caring for his jellyfish, he walks down the beach with a notebook, noting every organism that has washed ashore. It is a remarkable sight, the solitary figure in flip-flops, tramping pigeon-toed across the 400-yard length of the beach, hunched over, his floppy hair jogging in the breeze, as he intently scrutinizes the sand. He collates his data and publishes it in papers with titles like "Stranding Records of Fishes on Kitahama Beach" and "The First Occurrence of Bythotiara Species in Tanabe Bay." He is an active member of a dozen scientific societies and writes a jellyfish-of-the-week column in the local newspaper. Kubota says he has introduced his readers to more than 100 jellyfish so far.

Given Kubota's obsessive focus on his work, it is not surprising that he has been forced to neglect other areas of his life. He never cooks and tends to bring takeout to his office. At the lab, he wears T-shirts—bearing images of jellyfish—and sweatpants. He is overdue for a haircut. And his office is a mess. It does not appear to have been organized since he began nurturing his *Turritopsis*. The door opens just wide enough to admit a man of Kubota's stature. It is blocked from opening farther by a chest-high cabinet, on the surface of which are balanced several hundred objects Kubota has retrieved from beaches—seashells, bird feathers, crab claws, and desiccated coral. The desk is invisible beneath a stack of opened books. Fifty toothbrushes are crammed into a cup on the rusting aluminum sink. There are framed pictures on the wall, most of them depicting jellyfish, including one childish drawing done in crayons. I asked Kubota, who has two adult sons, whether one of his children had made it. He laughed, shaking his head.

"I'm not a very good artist," he said. I followed his glance to his desk, where there was a box of crayons.

The bookshelves that line the walls are jammed to overflowing with textbooks, journals, and science books, as well as a number of titles in English: Frank Herbert's *Dune, The Works of Aristotle, The Life and Death of Charles Darwin.* Kubota first read Darwin's *On the Origin of Species* in high school. It was one of the formative experiences of his life; before that, he thought he would grow up to be an archaeologist. He was then already fascinated with what he calls the "mystery of human life"—where did we come from and why? —and hoped that in the ancient civilizations he might discover the answers he sought. But after reading Darwin, he realized that he would have to look deeper into the past, beyond the dawn of human existence.

Kubota grew up in Matsuyama, on the southern island of Shikoku. Though his father was a teacher, Kubota didn't get excellent marks at his high school, where he was a generation behind Kenzaburo Oe. "I didn't study," he said. "I only read science fiction." But when he was admitted to college, his grandfather bought him a biological encyclopedia. It sits on one of his office shelves, beside a sepia-toned portrait of his grandfather.

"I learned a lot from that book," Kubota said. "I read every page." He was especially impressed by the phylogenetic tree, the taxonomic diagram that Darwin called the Tree of Life. Darwin included one of the earliest examples of a Tree of Life in *On the Origin of Species*—it is the book's only illustration. Today the outermost twigs and buds of the Tree of Life are occupied by mammals and birds, while at the base of the trunk lie the most primitive phyla—Porifera (sponges), Platyhelminthes (flatworms), Cnidaria (jellyfish).

"The mystery of life is not concealed in the higher animals," Kubota told me. "It is concealed in the root. And at the root of the Tree of Life is the jellyfish."

Until recently, the notion that human beings might have anything of value to learn from a jellyfish would have been considered absurd. Your typical cnidarian does not, after all, appear to have much in common with a human being. It has no brain, for instance, and no heart. It has a single orifice through which its food and waste pass—it eats, in other words, out of its own anus. But the Human Genome Project, completed in 2003, suggested otherwise. Though it had been estimated that our genome contained more than 100,000 protein-coding genes, it turned out that the number was closer to 21,000. This meant we had about the same number of genes as chickens, roundworms, and fruit flies. In a separate study, published in 2005, cnidarians were found to have a much more complex genome than previously imagined.

"There's a shocking amount of genetic similarity between jel-

lyfish and human beings," said Kevin J. Peterson, a molecular pa-
leobiologist who contributed to that study, when I visited him at
his Dartmouth office. From a genetic perspective, apart from the
fact that we have two genome duplications, "we look like a damn
jellyfish."

This may have implications for medicine, particularly the fields
of cancer research and longevity. Peterson is now studying micro-
RNAs (commonly denoted as miRNA), tiny strands of genetic material that regulate gene expression. MiRNA act as an on-off switch
for genes. When the switch is off, the cell remains in its primitive,
undifferentiated state. When the switch turns on, a cell assumes its
mature form: it can become a skin cell, for instance, or a tentacle
cell. MiRNA also serve a crucial role in stem-cell research—they
are the mechanism by which stem cells differentiate. Most cancers,
we have recently learned, are marked by alterations in miRNA.
Researchers even suspect that alterations in miRNA may be a *cause*
of cancer. If you turn a cell's miRNA "off," the cell loses its identity
and begins acting chaotically—it becomes, in other words, cancer-
ous.

Hydrozoans provide an ideal opportunity to study the behavior of miRNA for two reasons. They are extremely simple organisms, and miRNA are crucial to their biological development. But
because there are so few hydroid experts, our understanding of
these species is staggeringly incomplete.

"Immortality might be much more common than we think," Peterson said. "There are sponges out there that we know have been
there for decades. Sea-urchin larvae are able to regenerate and
continuously give rise to new adults." He continued: "This might
be a general feature of these animals. They never really die."

Peterson is closely following the work of Daniel Martinez, a biologist at Pomona College and one of the world's leading hydroid
scholars. The National Institutes of Health has awarded Martínez
a five-year, $1.26 million research grant to study the hydra—a spe-
cies that resembles a polyp but never yields medusas. Its body is
almost entirely composed of stem cells that allow it to regenerate
itself continuously. As a PhD candidate, Martinez set out to prove
that hydra were mortal. But his research of the last fifteen years
has convinced him that hydra can, in fact, survive forever and are
"truly immortal."

"It's important to keep in mind that we're not dealing with

something that's completely different from us," Martínez told me. "Genetically hydra are the same as human beings. We're variations of the same theme."

As Peterson told me: "If I studied cancer, the last thing I would study is cancer, if you take my point. I would not be studying thyroid tumors in mice. I'd be working on hydra."

Hydrozoans, he suggests, may have made a devil's bargain. In exchange for simplicity—no head or tail, no vision, eating out of their own anus—they gained immortality. These peculiar, simple species may represent an opportunity to learn how to fight cancer, old age, and death.

But most hydroid experts find it nearly impossible to secure financing. "Who's going to take a chance on a scientist who doesn't work on mammals, let alone a jellyfish?" Peterson said. "The granting agencies are always talking about trying to be imaginative and reinvigorate themselves, but of course you're stuck in a lot of bureaucracy . . . The pie is only so big."

Even some of Kubota's peers are cautious when speaking about potential medical applications in *Turritopsis* research. "It is difficult to foresee how much and how fast . . . *Turritopsis dohrnii* can be useful to fight diseases," Stefano Piraino, a colleague of Ferdinando Boero's, told me in an e-mail. "Increasing human longevity has no meaning, it is ecological nonsense. What we may expect and work on is to improve the quality of life in our final stages."

Martínez says that hydra, the species he studies, is more promising. "*Turritopsis* is cool," he told me. "Don't get me wrong. It's interesting that it does this weird, peculiar thing, and I support researching it further, but I don't think it's going to teach us a lot about human beings."

Kubota sees it differently. "The immortal medusa is the most miraculous species in the entire animal kingdom," he said. "I believe it will be easy to solve the mystery of immortality and apply ultimate life to human beings."

Kubota can be encouraged by the fact that many of the greatest advancements in human medicine came from observations made about animals that, at the time, seemed to have little or no resemblance to man. In eighteenth-century England, observation of dairymaids exposed to cowpox helped establish that the disease inoculated them against smallpox; the bacteriologist Alexander Fleming accidentally discovered penicillin when one of his petri dishes grew a mold; and, most recently, scientists in Wyoming studying nematode worms found genes similar to those inactivated by cancer in humans, leading them to believe that they could be a target for new cancer drugs. One of the Wyoming researchers said in a news release that they hoped those genes could "contribute to the arsenal of diverse therapeutic approaches used to treat and cure many types of cancer."

And so Kubota continues to accumulate data on his own simple organism every day of his life.

There is a second photograph on Shin Kubota's office shelf, beside the portrait of his grandfather. It shows a class of young university students posing on the campus of Ehime University, in Matsuyama. The photograph is forty years old, but the twenty-year-old Kubota is immediately recognizable—the round face, the smiling eyes, the floppy black hair. He sighed when I asked him about it.

"So young then," he said. "So old now."

I told him that he didn't look very different from the young man in the picture. He's perhaps a few pounds heavier, and though his features are not quite as boyish, he retains the exuberant energy of a middle-schooler, and his hair is naturally jet black. Yes, he said, but his hair hasn't always been black. He explained that five years ago, when he turned fifty-five, he experienced what he called a scare.

It was a stressful time for Kubota. He had separated from his wife, his children had moved out of the house, his eyesight was fading, and he had begun to lose his hair. It was particularly noticeable around his temples. He blames his glasses, which he wore on a band around his head. He needed them to write but not for the microscope, so every time he raised or lowered his glasses, the band wore away the hair at his temples. When the hair grew back, it came in white. He felt as if he had aged a lifetime in one year. "It was very astonishing for me," he said. "I had become old."

I told him that he looked much better now—significantly younger than his age.

"Too old," he said, scowling. "I want to be young again. I want to become miracle immortal man."

As if to distract himself from this trajectory of thought, he removed a petri cup from his refrigerator unit. He held it under the light so I could see the ghostly *Turritopsis* suspended within. It was still, waiting.

"Watch," he said. "I will make this medusa rejuvenate."

The most reliable way to make the immortal jellyfish age in reverse, Kubota explained to me, is to mutilate it. With two fine metal picks, he began to perforate the medusa's mesoglea, the gelatinous tissue that composes the bell. After Kubota poked it six times, the medusa behaved like any stabbing victim—it lay on its side and began twitching spasmodically. Its tentacles stopped undulating, and its bell slightly puckered. But Kubota, in what appeared a misdirected act of sadism, didn't stop there. He stabbed it fifty times in all. The medusa had long since stopped moving. It lay limp, crippled, its mesoglea torn, the bell deflated. Kubota looked satisfied.

"You rejuvenate!" he yelled at the jellyfish. Then he started laughing.

We checked on the stab victim every day that week to watch its transformation. On the second day, the depleted, gelatinous mess had attached itself to the floor of the petri dish; its tentacles were bent in on themselves. "It's transdifferentiating," Kubota said. "Dynamic changes are occurring." By the fourth day the tentacles were gone, and the organism ceased to resemble a medusa entirely; it looked instead like an amoeba. Kubota called this a "meatball." By the end of the week, stolons had begun to shoot out of the meatball.

This method is, in a certain sense, cheating, as physical distress induces rejuvenation. But the process also occurs naturally when the medusa grows old or sick. In Kubota's most recent paper on *Turritopsis*, he documented the natural rejuvenation of a single colony in his lab between 2009 and 2011. The idea was to see how quickly the species would regenerate itself when left to its own devices. During the two-year period, the colony rebirthed itself ten times, in intervals as brief as one month. In his paper's conclusion, published in the journal *Biogeography*, Kubota wrote, "Turritopsis will be kept forever by the present method and will . . . contribute to any study for everyone in the future."

He has made other significant findings in recent years. He has learned, for instance, that certain conditions inhibit rejuvenation: starvation, large bell size, and water colder than 72 degrees. And he has made progress in solving the largest mystery of all. The secret of the species's immortality, Kubota now believes, is hidden in the tentacles. But he will need more financing for experiments, as well as assistance from a geneticist or a molecular biologist, to figure out how the immortal jellyfish pulls it off. Even so, he thinks we're close to solving the species's mystery—that it's a matter of years, perhaps a decade or two. "Human beings are so intelligent," he told me, as if to reassure me. But then he added a caveat. "Before we achieve immortality," he said, "we must evolve first. The heart is not good."

I assumed that he was making a biological argument—that the organ is not biologically capable of infinite life, that we needed to design new, artificial hearts for longer, artificial lives. But then I realized that he wasn't speaking literally. By heart he meant the human spirit.

"Human beings must learn to love nature," he said. "Today the countryside is obsolete. In Japan, it has disappeared. Big metropolitan places have appeared everywhere. We are in the garbage. If this continues, nature will die."

Man, he explained, is intelligent enough to achieve biological immortality. But we don't deserve it. This sentiment surprised me coming from a man who has dedicated his life to pursuing immortality.

"Self-control is very difficult for humans," he continued. "In order to solve this problem, spiritual change is needed."

This is why, in the years since his "scare," Kubota has begun a second career. In addition to being a researcher, professor, and guest speaker, he is now a songwriter. Kubota's songs have been featured on national television, are available on karaoke machines across Japan, and have made him a minor Japanese celebrity—the Japanese equivalent of Bill Nye the Science Guy.

It helps that in Japan, the nation with the world's oldest population, the immortal jellyfish has a relatively exalted status in popular culture. Its reputation was boosted in 2003 by a television drama, *Fourteen Months*, in which the heroine takes a potion, extracted from the immortal jellyfish, that causes her to age in reverse. Since then Kubota has appeared regularly on television and radio shows. He showed me recent clips from his television reel and translated them for me. In March *Morning No. 1*, a Japanese morning show, devoted an episode to Shirahama. After a segment on the onsen, the hosts visited Kubota at the Seto Aquarium, where he talked about *Turritopsis*. "I want to become young, too!" one host shrieked. On *Love Laboratory,* a science show, Kubota discussed his recent experiments while collecting samples on the Shirahama wharf. "I envy the immortal medusa!" gushed the hostess. On *Feeding Our Bodies,* a similar program, Kubota addressed the camera: "Among the animals, the immortal jellyfish is the most splendid." There followed an interview with 100-year-old twins.

But no television appearance is complete without a song. For his performances, he transforms himself from Dr. Shin Kubota, erudite marine biologist in jacket and tie, into Mr. Immortal Jellyfish Man. His superhero alter ego has its own costume: a white lab jacket, scarlet-red gloves, red sunglasses, and a red rubber hat, designed to resemble a medusa, with dangling rubber tentacles. With help from one of his sons, an aspiring musician, Kubota has written dozens of songs in the last five years and released six albums. Many of his songs are odes to *Turritopsis*. These include "I Am Scarlet Medusa," "Life Forever," "Scarlet Medusa—an Eternal Witness," "Die-Hard Medusa," and his catchiest number, "Scarlet Medusa Chorus."

> My name is Scarlet Medusa,
> A teeny tiny jellyfish.
> But I have a special secret
> that no others may possess
> I can—yes, I can!—rejuvenate.

Other songs apotheosize different forms of marine life: "We Are the Sponges—a Song of the Porifera," "Viva! Variety Cnidaria," and "Poking Diving Horsehair Worm Mambo." There is also "I Am Shin Kubota."

> My name is Shin Kubota
> Associate professor of Kyoto University
> At Shirahama, Wakayama Prefecture.
> I live next to an aquarium
> Enjoying marine-biology research.
> Every day, I walk on the beach
> Scooping up with a plankton net
> Searching for wondrous creatures
> Searching for unknown jellyfish.
> [I] dedicate my life to small creatures

Patrolling the beaches every day.
Hot spring sandals are always on,
Necessary item to get in the sea.
Scarlet medusa rejuvenates,
Scarlet medusa is immortal.

"He is important for the aquarium," Akira Asakura, the Seto Lab director told me. "People come because they see him on television and become interested in the immortal medusa and marine life in general. He is a very good speaker, with a very wide range of knowledge."

Science classes regularly make field trips to meet Mr. Immortal Jellyfish Man. During my week in Shirahama, he was visited by a group of 150 ten- and eleven-year-olds who had prepared speeches and slide shows about *Turritopsis*. The group was too large to visit Seto, so they sat on the floor of a ballroom in a local hotel. After the children made their presentations ("I have jellyfish mania!" one girl exclaimed), Kubota took the stage. He spoke loudly, with great animation, calling on the children and peppering them with questions. How many species of animals are there on Earth? How many phyla are there? The karaoke video for "Scarlet Medusa Chorus" was projected on a large screen, and the giggling children sang along.

Kubota does not go to these lengths simply for his own amusement—though it is clear that he enjoys himself immensely. Nor does he consider his public educational work as secondary to his research. It is instead, he believes, the crux of his life's work.

"We must love plants—without plants we cannot live. We must love bacteria—without decomposition our bodies can't go back to the earth. If everyone learns to love living organisms, there will be no crime. No murder. No suicide. Spiritual change is needed. And the most simple way to achieve this is through song.

"Biology is specialized," he said, bringing his palms within inches of each other. "But songs?"

He spread his hands far apart, as if to indicate the size of the world.

Every night, once Kubota is finished with work, he grabs a bite to eat and heads to a karaoke bar. He sings karaoke for at least two hours a day. He owns a karaoke book that is 1,611 pages long, with dimensions somewhat larger than a phone book and even denser type. His goal is to sing at least one song from every page. Every time he sings a song, he underlines it in the book. Flipping through the volume, I saw that he had easily surpassed his goal.

"When I perform karaoke," he said, "another part of the brain is used. It's good to relax, to sing a heartfelt song. It's good to be loud."

His favorite karaoke bar is called Kibarashi, which translates loosely to "recreation" but literally means "fresh air." Kibarashi stands at the end of a residential street, away from the coastal road and the city's other main commercial stretches. He'd given me clear directions, but I struggled to find it. The street was silent and dark. I was ready to turn back, assuming I'd made a wrong turn, when I saw a small sign decorated with an illuminated microphone. When I opened the door, I found myself in what resembled a living room—couches, coffee tables, pots with plastic flowers, goldfish in small tanks. A low, narrow bar ran along one wall. A karaoke video of a tender Japanese ballad was playing on two televisions that hung from the ceiling. Kubota stood facing one of them, microphone in hand, swaying side to side, singing full-throatedly in his elegant mezzo-baritone. The bartender, a woman in her seventies, was seated behind the bar, tapping on her iPhone. Nobody else was there.

We sang for the next two hours—Elvis Presley, the Beatles, the Beastie Boys, and countless Japanese ballads and children's songs. At my request, Kubota sang his own songs, seven of which are listed in his karaoke book. Kibarashi's karaoke machine is part of an international network of karaoke machines, and the computer displays statistics for each song, including how many people in Japan have selected it in the past month. It seemed as if no one had selected Kubota's songs.

"Unfortunately they are not sung by many people," he told me. "They're not popular, because it's very difficult to love nature, to love animals."

On my last morning in Shirahama, Kubota called to cancel our final meeting. He had a bacterial infection in his eye and couldn't see clearly enough to look through his microscope. He was going to a specialist. He apologized repeatedly.

"Human beings very weak," he said, "Bacteria very strong. I want to be immortal!" He laughed his hearty laugh.

Turritopsis, it turns out, is also very weak. Despite being immortal, it is easily killed. *Turritopsis* polyps are largely defenseless against their predators, chief among them sea slugs. They can easily be suffocated by organic matter. "They're miracles of nature, but they're not complete," Kubota acknowledged. "They're still organisms. They're not holy. They're not God."

And their immortality is, to a certain degree, a question of semantics. "That word 'immortal' is distracting," says James Carlton, the professor of marine sciences at Williams. "If by 'immortal' you mean passing on your genes, then yes, it's immortal. But those are not the same cells anymore. The cells are immortal, but not necessarily the organism itself." To complete the Benjamin Button analogy, imagine the man, after returning to a fetus, being born again. The cells would be recycled, but the old Benjamin would be gone; in his place would be a different man with a new brain, a new heart, a new body. He would be a clone.

But we won't know for certain what this means for human beings until more research is done. That is the scientific method, after all: lost in the labyrinth, you must pursue every path, no matter how unlikely, or risk being devoured by the Minotaur. Kubota, for his part, fears that the lessons of the immortal jellyfish will be absorbed too soon, before man is ready to harness the science of immortality in an ethical manner. "We're very strange animals," he said. "We're so clever and civilized, but our hearts are very primitive. If our hearts weren't primitive, there wouldn't be wars. I'm worried that we will apply the science too early, like we did with the atomic bomb."

I remembered something he had said earlier in the week, when we were watching a music video for his song "Living Planet—Connections Between Forest, Sea and Rural Area." He described the song as an ode to the beauty of nature. The video was shot by his eighty-eight-year-old neighbor, a retired employee of Osaka Gas Company. Kubota's lyrics were superimposed over a sequence of images. There was Engetsu, its arch covered with moss and jutting oak and pine trees; craggy Mount Seppiko and gentle Mount Takane; the striated cliffs of Sandanbeki; the private beach at the

Seto Laboratory; a waterfall; a brook; a pond; and the cliff-side forests that abut the city, so dense and black that the trees seem to be secreting darkness.

"Nature is so beautiful," Kubota said, smiling wistfully. "If human beings disappeared, how peaceful it would be."

STEPHEN MARCHE

Is Facebook Making Us Lonely?

FROM *The Atlantic*

YVETTE VICKERS, A former *Playboy* playmate and B-movie star, best known for her role in *Attack of the 50-Foot Woman,* would have been eighty-three last August, but nobody knows exactly how old she was when she died. According to the Los Angeles coroner's report, she lay dead for the better part of a year before a neighbor and fellow actress, a woman named Susan Savage, noticed cobwebs and yellowing letters in her mailbox, reached through a broken window to unlock the door, and pushed her way through the piles of junk mail and mounds of clothing that barricaded the house. Upstairs she found Vickers's body, mummified, near a heater that was still running. Her computer was on too, its glow permeating the empty space.

The *Los Angeles Times* posted a story headlined MUMMIFIED BODY OF FORMER PLAYBOY PLAYMATE YVETTE VICKERS FOUND IN HER BENEDICT CANYON HOME, which quickly went viral. Within two weeks, by Technorati's count, Vickers's lonesome death was already the subject of 16,057 Facebook posts and 881 tweets. She had long been a horror-movie icon, a symbol of Hollywood's capacity to exploit our most basic fears in the silliest ways; now she was an icon of a new and different kind of horror: our growing fear of loneliness. Certainly she received much more attention in death than she did in the final years of her life. With no children, no religious group, and no immediate social circle of any kind, she had begun, as an elderly woman, to look elsewhere for companionship. Savage later told *Los Angeles* magazine that she had searched Vickers's phone bills for clues about the life that led to such an end. In the months before her grotesque death, Vickers had made calls not to friends or family but to distant fans who had found her through fan conventions and Internet sites.

Vickers's web of connections had grown broader but shallower, as has happened for many of us. We are living in an isolation that would have been unimaginable to our ancestors, and yet we have never been more accessible. Over the past three decades, technology has delivered to us a world in which we need not be out of contact for a fraction of a moment. In 2010, at a cost of $300 million, 800 miles of fiber-optic cable was laid between the Chicago Mercantile Exchange and the New York Stock Exchange to shave three milliseconds off trading times. Yet within this world of instant and absolute communication, unbounded by limits of time or space, we suffer from unprecedented alienation. We have never been more detached from one another, or lonelier. In a world consumed by ever more novel modes of socializing, we have less and less actual society. We live in an accelerating contradiction: the more connected we become, the lonelier we are. We were promised a global village; instead we inhabit the drab cul-de-sacs and endless freeways of a vast suburb of information.

At the forefront of all this unexpectedly lonely interactivity is Facebook, with 845 million users and $3.7 billion in revenue last year. The company hopes to raise $5 billion in an initial public offering later this spring, which will make it by far the largest Internet IPO in history. Some recent estimates put the company's potential value at $100 billion, which would make it larger than the global coffee industry—one addiction preparing to surpass the other. Facebook's scale and reach are hard to comprehend: last summer Facebook became, by some counts, the first website to receive 1 trillion page views in a month. In the last three months of 2011, users generated an average of 2.7 billion "likes" and comments every day. On whatever scale you care to judge Facebook —as a company, as a culture, as a country—it is vast beyond imagination.

Despite its immense popularity, or more likely because of it, Facebook has, from the beginning, been under something of a cloud of suspicion. The depiction of Mark Zuckerberg in *The Social Network* as a bastard with symptoms of Asperger's syndrome was nonsense. But it felt true. It felt true to Facebook, if not to Zuckerberg. The film's most indelible scene, the one that may well have earned it an Oscar, was the final, silent shot of an anomic Zuckerberg sending out a friend request to his ex-girlfriend, then waiting and clicking and waiting and clicking—a moment of superconnected loneliness preserved in amber. We have all been in that scene: transfixed by the glare of a screen, hungering for a response.

When you sign up for Google+ and set up your Friends circle, the program specifies that you should include only "your real friends, the ones you feel comfortable sharing private details with." That one little phrase, *your real friends*—so quaint, so charmingly mothering—perfectly encapsulates the anxieties that social media have produced: the fears that Facebook is interfering with our real friendships, distancing us from each other, making us lonelier; and that social networking might be spreading the very isolation it seemed designed to conquer.

Facebook arrived in the middle of a dramatic increase in the quantity and intensity of human loneliness, a rise that initially made the site's promise of greater connection seem deeply attractive. Americans are more solitary than ever before. In 1950 less than 10 percent of American households contained only one person. By 2010 nearly 27 percent of households had just one person. Solitary living does not guarantee a life of unhappiness, of course. In his recent book about the trend toward living alone, Eric Klinenberg, a sociologist at NYU, writes: "Reams of published research show that it's the quality, not the quantity of social interaction, that best predicts loneliness." True. But before we begin the fantasies of happily eccentric singledom, of divorcées dropping by their knitting circles after work for glasses of Drew Barrymore pinot grigio, or recent college graduates with perfectly articulated, steampunk-themed, 300-square-foot apartments organizing croquet matches with their book clubs, we should recognize that it is not just isolation that is rising sharply. It's loneliness, too. And loneliness makes us miserable.

We know intuitively that loneliness and being alone are not the same thing. Solitude can be lovely. Crowded parties can be agony. We also know, thanks to a growing body of research on the topic, that loneliness is not a matter of external conditions; it is a psychological state. A 2005 analysis of data from a longitudinal study of Dutch twins showed that the tendency toward loneliness has roughly the same genetic component as other psychological problems such as neuroticism or anxiety.

Still, loneliness is slippery, a difficult state to define or diagnose. The best tool yet developed for measuring the condition is the UCLA Loneliness Scale, a series of twenty questions that all begin with this formulation: "How often do you feel . . . ?" As in "How often do you feel that you are 'in tune' with the people around you?" And "How often do you feel that you lack companionship?" Measuring the condition in these terms, various studies have shown loneliness rising drastically over a very short period of recent history. A 2010 AARP survey found that 35 percent of adults older than forty-five were chronically lonely, as opposed to 20 percent of a similar group only a decade earlier. According to a major study by a leading scholar of the subject, roughly 20 percent of Americans—about 60 million people—are unhappy with their lives because of loneliness. Across the Western world, physicians and nurses have begun to speak openly of an epidemic of loneliness.

The new studies on loneliness are beginning to yield some surprising preliminary findings about its mechanisms. Almost every factor that one might assume affects loneliness does so only some of the time, and only under certain circumstances. People who are married are less lonely than single people, one journal article suggests, but only if their spouses are confidants. If one's spouse is not a confidant, marriage may not decrease loneliness. A belief in God might help, or it might not, as a 1990 German study comparing levels of religious feeling and levels of loneliness discovered. Active believers who saw God as abstract and helpful rather than as a wrathful, immediate presence were less lonely. "The mere belief in God," the researchers concluded, "was relatively independent of loneliness."

But it is clear that social interaction matters. Loneliness and being alone are not the same thing, but both are on the rise. We meet fewer people. We gather less. And when we gather, our bonds are less meaningful and less easy. The decrease in confidants—that is, in quality social connections—has been dramatic over the past twenty-five years. In one survey, the mean size of networks of personal confidants decreased from 2.94 people in 1985 to 2.08

in 2004. Similarly, in 1985, only 10 percent of Americans said they had no one with whom to discuss important matters, and 15 percent said they had only one such good friend. By 2004, 25 percent had nobody to talk to, and 20 percent had only one confidant.

In the face of this social disintegration, we have essentially hired an army of replacement confidants, an entire class of professional carers. As Ronald Dworkin pointed out in a 2010 paper for the Hoover Institution, in the late 1940s, the United States was home to 2,500 clinical psychologists, 30,000 social workers, and fewer than 500 marriage and family therapists. As of 2010, the country had 77,000 clinical psychologists, 192,000 clinical social workers, 400,000 nonclinical social workers, 50,000 marriage and family therapists, 105,000 mental-health counselors, 220,000 substance-abuse counselors, 17,000 nurse psychotherapists, and 30,000 life coaches. The majority of patients in therapy do not warrant a psychiatric diagnosis. This raft of psychic servants is helping us through what used to be called regular problems. We have outsourced the work of everyday caring.

We need professional carers more and more, because the threat of societal breakdown, once principally a matter of nostalgic lament, has morphed into an issue of public health. Being lonely is extremely bad for your health. If you're lonely, you're more likely to be put in a geriatric home at an earlier age than a similar person who isn't lonely. You're less likely to exercise. You're more likely to be obese. You're less likely to survive a serious operation and more likely to have hormonal imbalances. You are at greater risk of inflammation. Your memory may be worse. You are more likely to be depressed, to sleep badly, and to suffer dementia and general cognitive decline. Loneliness may not have killed Yvette Vickers, but it has been linked to a greater probability of having the kind of heart condition that did kill her.

And yet, despite its deleterious effect on health, loneliness is one of the first things ordinary Americans spend their money achieving. With money, you flee the cramped city to a house in the suburbs or, if you can afford it, a McMansion in the exurbs, inevitably spending more time in your car. Loneliness is at the American core, a byproduct of a long-standing national appetite for independence: the Pilgrims who left Europe willingly abandoned the bonds and strictures of a society that could not accept their right to be different. They did not seek out loneliness, but they accepted it as the price of their autonomy. The cowboys who set off to explore a seemingly endless frontier likewise traded away personal ties in favor of pride and self-respect. The ultimate American icon is the astronaut. Who is more heroic, or more alone? The price of self-determination and self-reliance has often been loneliness. But Americans have always been willing to pay that price.

Today the one common feature in American secular culture is its celebration of the self that breaks away from the constrictions of the family and the state and, in its greatest expressions, from all limits entirely. The great American poem is Whitman's "Song of Myself." The great American essay is Emerson's "Self-Reliance." The great American novel is Melville's *Moby-Dick,* the tale of a man on a quest so lonely that it is incomprehensible to those around him. American culture, high and low, is about self-expression and personal authenticity. Franklin Delano Roosevelt called individualism "the great watchword of American life."

Self-invention is only half of the American story, however. The drive for isolation has always been in tension with the impulse to cluster in communities that cling and suffocate. The Pilgrims, while fomenting spiritual rebellion, also enforced ferocious cohesion. The Salem witch trials, in hindsight, read like attempts to impose solidarity—as do the McCarthy hearings. The history of the United States is like the famous parable of the porcupines in the cold, from Schopenhauer's *Studies in Pessimism*—the ones who huddle together for warmth and shuffle away in pain, always separating and congregating.

We are now in the middle of a long period of shuffling away. In his 2000 book *Bowling Alone,* Robert D. Putnam attributed the dramatic postwar decline of social capital—the strength and value of interpersonal networks—to numerous interconnected trends in American life: suburban sprawl, television's dominance over culture, the self-absorption of the baby boomers, the disintegration of the traditional family. The trends he observed continued through the prosperity of the aughts and have only become more pronounced with time: the rate of union membership declined in 2011 again; screen time rose; the Masons and the Elks continued their slide into irrelevance. We are lonely because we want to be lonely. We have made ourselves lonely.

The question of the future is this: Is Facebook part of the separating or part of the congregating; is it a huddling-together for warmth or a shuffling-away in pain?

Well before Facebook, digital technology was enabling our tendency for isolation to an unprecedented degree. Back in the 1990s, scholars started calling the contradiction between an increased opportunity to connect and a lack of human contact the "Internet paradox." A prominent 1998 article on the phenomenon by a team of researchers at Carnegie Mellon showed that increased Internet usage was already coinciding with increased loneliness. Critics of the study pointed out that the two groups that participated in the study—high-school journalism students who were heading to university and socially active members of community development boards—were statistically likely to become lonelier over time. Which brings us to a more fundamental question: Does the Internet make people lonely, or are lonely people more attracted to the Internet?

The question has intensified in the Facebook era. A recent study out of Australia (where close to half the population is active on Facebook), titled "Who Uses Facebook?" found a complex and sometimes confounding relationship between loneliness and social networking. Facebook users had slightly lower levels of "social loneliness"—the sense of not feeling bonded with friends—but "significantly higher levels of family loneliness"—the sense of not feeling bonded with family. It may be that Facebook encourages more contact with people outside of our household, at the expense of our family relationships—or it may be that people who have unhappy family relationships in the first place seek companionship through other means, including Facebook. The researchers also found that lonely people are inclined to spend more time on Facebook: "One of the most noteworthy findings," they wrote, "was the tendency for neurotic and lonely individuals to spend greater amounts of time on Facebook per day than non-lonely individuals." And they found that neurotics are more likely to prefer to use the wall, while extroverts tend to use chat features in addition to the wall.

Moira Burke, until recently a graduate student at the Human-Computer Institute at Carnegie Mellon, used to run a longitudinal study of 1,200 Facebook users. That study, which is ongoing, is one of the first to step outside the realm of self-selected college students and examine the effects of Facebook on a broader population over time. She concludes that the effect of Facebook depends on what you bring to it. Just as your mother said: you get out only what you put in. If you use Facebook to communicate directly with other individuals—by using the "like" button, commenting on friends' posts, and so on—it can increase your social capital. Personalized messages, or what Burke calls "composed communication," are more satisfying than "one-click communication"—the lazy click of a like. "People who received composed communication became less lonely, while people who received one-click communication experienced no change in loneliness," Burke tells me. So you should inform your friend in writing how charming her son looks with Harry Potter cake smeared all over his face, and how interesting her sepia-toned photograph of that tree-framed bit of skyline is, and how cool it is that she's at whatever concert she happens to be at. That's what we all want to hear. Even better than sending a private Facebook message is the semipublic conversation, the kind of back-and-forth in which you half ignore the other people who may be listening in. "People whose friends write to them semipublicly on Facebook experience decreases in loneliness," Burke says.

On the other hand, nonpersonalized use of Facebook—scanning your friends' status updates and updating the world on your own activities via your wall, or what Burke calls "passive consumption" and "broadcasting"—correlates to feelings of disconnectedness. It's a lonely business, wandering the labyrinths of our friends' and pseudo-friends' projected identities, trying to figure out what part of ourselves we ought to project, who will listen, and what they will hear. According to Burke, passive consumption of Facebook also correlates to a marginal increase in depression. "If two women each talk to their friends the same amount of time, but one of them spends more time reading about friends on Facebook as well, the one reading tends to grow slightly more depressed," Burke says. Her conclusion suggests that my sometimes unhappy reactions to Facebook may be more universal than I had realized. When I scroll through page after page of my friends' descriptions of how accidentally eloquent their kids are, and how their husbands are endearingly bumbling, and how they're all about to eat a home-cooked meal prepared with fresh local organic produce bought at the farmers' market and then go for a jog and maybe check in at the office because they're so busy getting ready to hop on a plane for a week of luxury dogsledding in Lapland, I do grow slightly more miserable. A lot of other people doing the same thing feel a little bit worse, too.

Still, Burke's research does not support the assertion that Facebook creates loneliness. The people who experience loneliness on Facebook are lonely away from Facebook too, she points out; on Facebook, as everywhere else, correlation is not causation. The popular kids are popular, and the lonely skulkers skulk alone. Perhaps it says something about me that I think Facebook is primarily a platform for lonely skulking. I mention to Burke the widely reported study, conducted by a Stanford graduate student, that showed how believing that others have strong social networks can lead to feelings of depression. What does Facebook communicate, if not the impression of social bounty? Everybody else looks so happy on Facebook, with so many friends, that our own social networks feel emptier than ever in comparison. Doesn't that *make* people feel lonely? "If people are reading about lives that are much better than theirs, two things can happen," Burke tells me. "They can feel worse about themselves, or they can feel motivated."

Burke will start working at Facebook as a data scientist this year.

John Cacioppo, the director of the Center for Cognitive and Social Neuroscience at the University of Chicago, is the world's leading expert on loneliness. In his landmark book *Loneliness*, released in 2008, he revealed just how profoundly the epidemic of loneliness is affecting the basic functions of human physiology. He found higher levels of epinephrine, the stress hormone, in the morning urine of lonely people. Loneliness burrows deep: "When we drew blood from our older adults and analyzed their white cells," he writes, "we found that loneliness somehow penetrated the deepest recesses of the cell to alter the way genes were being expressed." Loneliness affects not only the brain, then, but the basic process of DNA transcription. When you are lonely, your whole body is lonely.

To Cacioppo, Internet communication allows only ersatz intimacy. "Forming connections with pets or online friends or even God is a noble attempt by an obligatorily gregarious creature to satisfy a compelling need," he writes. "But surrogates can never make up completely for the absence of the real thing." The "real thing" being actual people, in the flesh. When I speak to Cacioppo, he is refreshingly clear on what he sees as Facebook's effect on society. Yes, he allows, some research has suggested that the greater the number of Facebook friends a person has, the less lonely she is. But he argues that the impression this creates can be misleading. "For the most part," he says, "people are bringing their old friends, and feelings of loneliness or connectedness, to Facebook." The idea that a website could deliver a more friendly, interconnected world is bogus. The depth of one's social network outside Facebook is what determines the depth of one's social network within Facebook, not the other way around. Using social media doesn't create new social networks; it just transfers established networks from one platform to another. For the most part, Facebook doesn't destroy friendships—but it doesn't create them, either.

In one experiment, Cacioppo looked for a connection between the loneliness of subjects and the relative frequency of their interactions via Facebook, chat rooms, online games, dating sites, and face-to-face contact. The results were unequivocal. "The greater the proportion of face-to-face interactions, the less lonely you are," he says. "The greater the proportion of online interactions, the lonelier you are." Surely, I suggest to Cacioppo, this means that Facebook and the like inevitably make people lonelier. He disagrees. Facebook is merely a tool, he says, and like any tool, its effectiveness will depend on its user. "If you use Facebook to increase face-to-face contact," he says, "it increases social capital." So if social media let you organize a game of football among your friends, that's healthy. If you turn to social media instead of playing football, however, that's unhealthy.

"Facebook can be terrific, if we use it properly," Cacioppo continues. "It's like a car. You can drive it to pick up your friends. Or you can drive alone." But hasn't the car increased loneliness? If cars created the suburbs, surely they also created isolation. "That's because of how we use cars," Cacioppo replies. "How we use these technologies can lead to more integration rather than more isolation."

The problem, then, is that we invite loneliness, even though it makes us miserable. The history of our use of technology is a history of isolation desired and achieved. When the Great Atlantic and Pacific Tea Company opened its A&P stores, giving Americans

self-service access to groceries, customers stopped having relationships with their grocers. When the telephone arrived, people
stopped knocking on their neighbors' doors. Social media bring
this process to a much wider set of relationships. Researchers at
the HP Social Computing Lab who studied the nature of people's
connections on Twitter came to a depressing, if not surprising,
conclusion: "Most of the links declared within Twitter were meaningless from an interaction point of view." I have to wonder: What
other point of view is meaningful?

Loneliness is certainly not something that Facebook or Twitter or
any of the lesser forms of social media is doing to us. We are doing it to ourselves. Casting technology as some vague, impersonal
spirit of history forcing our actions is a weak excuse. We make
decisions about how we use our machines, not the other way
around. Every time I shop at my local grocery store, I am faced
with a choice. I can buy my groceries from a human being or from
a machine. I always, without exception, choose the machine. It's
faster and more efficient, I tell myself, but the truth is that I prefer not having to wait with the other customers who are lined up
alongside the conveyor belt: the hipster mom who disapproves of
my high-carbon-footprint pineapple; the lady who tenses to the
point of tears while she waits to see if the gods of the credit-card
machine will accept or decline; the old man whose clumsy feeble-
ness requires a patience that I don't possess. Much better to bypass
the whole circus and just ring up the groceries myself.
     Our omnipresent new technologies lure us toward increasingly
superficial connections at exactly the same moment that they
make avoiding the mess of human interaction easy. The beauty of
Facebook, the source of its power, is that it enables us to be social
while sparing us the embarrassing reality of society—the accidental revelations we make at parties, the awkward pauses, the farting
and the spilled drinks and the general gaucherie of face-to-face
contact. Instead, we have the lovely smoothness of a seemingly social machine. Everything's so simple: status updates, pictures, your
wall.
     But the price of this smooth sociability is a constant compul-
sion to assert one's own happiness, one's own fulfillment. Not only
must we contend with the social bounty of others; we must fos-

ter the appearance of our own social bounty. Being happy all the time, pretending to be happy, actually attempting to be happy—it's exhausting. Last year a team of researchers led by Iris Mauss at the University of Denver published a study looking into "the paradoxical effects of valuing happiness." Most goals in life show a direct correlation between valuation and achievement. Studies have found, for example, that students who value good grades tend to have higher grades than those who don't value them. Happiness is an exception. The study came to a disturbing conclusion:

> Valuing happiness is not necessarily linked to greater happiness. In fact, under certain conditions, the opposite is true. Under conditions of low (but not high) life stress, the more people valued happiness, the lower were their hedonic balance, psychological well-being, and life satisfaction, and the higher their depression symptoms.

The more you try to be happy, the less happy you are. Sophocles made roughly the same point.

Facebook, of course, puts the pursuit of happiness front and center in our digital life. Its capacity to redefine our very concepts of identity and personal fulfillment is much more worrisome than the data mining and privacy practices that have aroused anxieties about the company. Two of the most compelling critics of Facebook—neither of them a Luddite—concentrate on exactly this point. Jaron Lanier, the author of *You Are Not a Gadget,* was one of the inventors of virtual-reality technology. His view of where social media are taking us reads like dystopian science fiction: "I fear that we are beginning to design ourselves to suit digital models of us, and I worry about a leaching of empathy and humanity in that process." Lanier argues that Facebook imprisons us in the business of self-presenting, and this, to his mind, is the site's crucial and fatally unacceptable downside.

Sherry Turkle, a professor of computer culture at MIT who in 1995 published the digital-positive analysis *Life on the Screen,* is much more skeptical about the effects of online society in her 2011 book, *Alone Together:* "These days, insecure in our relationships and anxious about intimacy, we look to technology for ways to be in relationships and protect ourselves from them at the same time." The problem with digital intimacy is that it is ultimately incomplete: "The ties we form through the Internet are not, in the end, the ties that bind. But they are the ties that preoccupy," she writes. "We don't want to intrude on each other, so instead we constantly intrude on each other, but not in 'real time.'"

Lanier and Turkle are right, at least in their diagnoses. Self-presentation on Facebook is continuous, intensely mediated, and possessed of a phony nonchalance that eliminates even the potential for spontaneity. ("Look how casually I threw up these three photos from the party at which I took 300 photos!") Curating the exhibition of the self has become a 24/7 occupation. Perhaps not surprisingly, then, the Australian study "Who Uses Facebook?" found a significant correlation between Facebook use and narcissism: "Facebook users have higher levels of total narcissism, exhibitionism, and leadership than Facebook nonusers," the study's authors wrote. "In fact, it could be argued that Facebook specifically gratifies the narcissistic individual's need to engage in self-promoting and superficial behavior."

Rising narcissism isn't so much a trend as the trend behind all other trends. In preparation for the 2013 edition of its diagnostic manual, the psychiatric profession is currently struggling to update its definition of narcissistic personality disorder. Still, generally speaking, practitioners agree that narcissism manifests in patterns of fantastic grandiosity, craving for attention, and lack of empathy. In a 2008 survey, 35,000 American respondents were asked if they had ever had certain symptoms of narcissistic personality disorder. Among people older than sixty-five, 3 percent reported symptoms. Among people in their twenties, the proportion was nearly 10 percent. Across all age groups, one in sixteen Americans has experienced some symptoms of NPD. And loneliness and narcissism are intimately connected: a longitudinal study of Swedish women demonstrated a strong link between levels of narcissism in youth and levels of loneliness in old age. The connection is fundamental. Narcissism is the flip side of loneliness, and either condition is a fighting retreat from the messy reality of other people.

A considerable part of Facebook's appeal stems from its miraculous fusion of distance with intimacy, or the illusion of distance with the illusion of intimacy. Our online communities become engines of self-image, and self-image becomes the engine of community. The real danger with Facebook is not that it allows us to isolate ourselves, but that by mixing our appetite for isolation with our vanity, it threatens to alter the very nature of solitude. The new isolation is not of the kind that Americans once idealized, the lonesomeness of the proudly nonconformist, independent-minded, solitary stoic, or that of the astronaut who blasts into new worlds. Facebook's isolation is a grind. What's truly stagger-ing about Facebook usage is not its volume — 750 million photo-graphs uploaded over a single weekend — but the constancy of the performance it demands. More than half its users — and one of every thirteen people on Earth is a Facebook user — log on every day. Among eighteen- to thirty-four-year-olds, nearly half check Facebook minutes after waking up, and 28 percent do so before getting out of bed. The relentlessness is what is so new, so poten-tially transformative. Facebook never takes a break. We never take a break. Human beings have always created elaborate acts of self-presentation. But not all the time, not every morning, before we even pour a cup of coffee. Yvette Vickers's computer was on when she died.

Nostalgia for the good old days of disconnection would not just be pointless, it would be hypocritical and ungrateful. But the very magic of the new machines, the efficiency and elegance with which they serve us, obscures what isn't being served: everything that matters. What Facebook has revealed about human nature — and this is not a minor revelation — is that a connection is not the same thing as a bond, and that instant and total connection is no salvation, no ticket to a happier, better world or a more liberated version of humanity. Solitude used to be good for self-reflection and self-reinvention. But now we are left thinking about who we are all the time, without ever really thinking about who we are. Facebook denies us a pleasure whose profundity we had under-estimated: the chance to forget about ourselves for a while, the chance to disconnect.

MARK BOWDEN

The Measured Man

FROM *The Atlantic*

LIKE MANY PEOPLE who are careful about their weight, Larry Smarr once spent two weeks measuring everything he put in his mouth. He charted each serving of food in grams or teaspoons and broke it down into these categories: protein, carbohydrates, fat, sodium, sugar, and fiber.

Larry used the data to fine-tune his diet. With input nailed down, he turned to output. He started charting the calories he burns, in workouts on an elliptical trainer and in the steps he takes each day. If the number on his pedometer falls short of his prescribed daily 7,000, he will find an excuse to go for a walk. Picture a tall, slender man with the supple, slightly deflated look of someone who has lost a lot of weight, plodding purposefully in soft shoes along the sunny sidewalks of La Jolla, California.

Of course, where outputs are concerned, calories are only part of the story, and it is here that Larry begins to differ from your typical health nut. Because human beings also produce waste products, foremost among them . . . well, poop. Larry collects his and has it analyzed. He is deep into the biochemistry of his feces, keeping detailed charts of their microbial contents. Larry has even been known to haul carefully boxed samples out of his kitchen refrigerator to show incautious visitors. He is eloquent on the subject. He could *sell* the stuff.

"Have you ever figured how information-rich your stool is?" Larry asks me with a wide smile, his gray-green eyes intent behind rimless glasses. "There are about one hundred billion bacteria per gram. Each bacterium has DNA whose length is typically one to ten megabases—call it one million bytes of information. This means human stool has a data capacity of one hundred thousand terabytes of information stored per gram. That's many orders of magnitude more information density than, say, in a chip in your smartphone or your personal computer. So your stool is far more interesting than a computer."

Larry's fascination is less with feces themselves than with the data they yield. He is not a doctor or a biochemist; he's a computer scientist—one of the early architects of the Internet, in fact. Today he directs a world-class research center on two University of California campuses, San Diego and Irvine, called the California Institute for Telecommunications and Information Technology, or "Calit2" (the *2* represents the repeated *I* and *T* initials). The future is arriving faster at Calit2 than it is in most places. Larry says his eyes are focused "ten years ahead," which in computer terms is more like a century or two, given how rapidly the machines are transforming modern life. Intent on that technological horizon, Larry envisions a coming revolution in medicine, and he is bringing his intellect and his institute to bear on it.

At sixty-three, he is engaged in a computer-aided study of the human body—specifically, *his* body. It's the start of a process that he believes will help lead, within ten years, to the development of "a distributed planetary computer of enormous power," one that is composed of a billion processors and will enable scientists to create, among many other things, a working computational model of your body. Your particular body, mind you, not just some generalized atlas of the human frame, but a working model of your unique corpus, grounded in your own genome, and—using data collected by nanosensors and transmitted by smartphone —refreshed continually with measurements from your body's insides. This information stream will be collated with similar readings from millions of other similarly monitored bodies all over the planet. Mining this enormous database, software will produce detailed guidance about diet, supplements, exercise, medication, or treatment—guidance based not on the current practice of lumping symptoms together into broad categories of disorders, but on a precise reading of your own body's peculiarities and its status in real time.

"And at that point," says Larry, in a typically bold pronouncement that would startle generations of white-coated researchers, "you now have, for the first time in history, a scientific basis for medicine."

When Socrates exhorted his followers, "Know thyself," he could not have imagined an acolyte so avid, or so literal, as Larry. You've heard of people who check their pulse every few minutes? Amateurs. When Larry works out, an armband records skin temperature, heat flux, galvanic skin response, and acceleration in three dimensions. When he sleeps, a headband monitors the patterns of his sleep every 30 seconds. He has his blood drawn as many as eight times a year and regularly tracks 100 separate markers. He is on a first-name basis with his ultrasound and MRI technicians, who provide him with 3-D images of his body, head to toe. Regular colonoscopies record the texture and color of his innards. And then there are the stool samples—last year Larry sent specimens to a lab for analysis nine times.

Larry is a mild, gentle soul, someone generally more interested in talking about you than about himself. He does not go out of his way to get your attention, and nothing about him is remotely annoying or evangelical. But if you show an interest in his project and start asking questions—look out. Beneath the calm and the deference, Larry is an intellectual pitchman of the first order. His quest to know burns with the pure intellectual passion of a precocious ten-year-old. He visibly shudders with pleasure at a good, hard question; his shoulders subtly rise and square, and his forehead leans into the task. Because Larry is on a mission. He's out to change the world and, along the way, defeat at least one incurable disease: his own. (More on this in a moment.)

Larry is in the vanguard of what some call the "quantified life," which envisions replacing the guesswork and supposition presently guiding individual health decisions with specific guidance tailored to the particular details of each person's body. Because of his accomplishments and stature in his field, Larry cannot easily be dismissed as a kook. He believes in immersing himself in his work. Years ago, at the University of Illinois, when he was taking part in an experiment to unravel complex environmental systems with supercomputers, Larry installed a coral-reef aquarium in his home, complete with shrimp and sixteen other phyla of small marine critters. It was maddeningly fragile. The coral kept peeling off the rocks and dying. He eventually discovered that just five drops of molybdenum, a metallic element, in a 250-gallon tank once a

week solved the problem. That such a tiny factor played so deci
a role helped him better grasp the complexity of the situati
And as he fought to sustain the delicate ecosystem in his tank, he
developed a personal feel for the larger problem his team was trying to solve.

Today he is preoccupied with his own ecosystem. The way a
computer scientist tends to see it, a genome is a given individual's
basic program. Mapping one used to cost billions. Today it can be
done for thousands, and soon the price will drop below $1,000.
Once people know their genetic codes and begin thoroughly
monitoring their bodily systems, they will theoretically approach
the point where computers can "know" a lot more about them
than any doctor ever could. In such a world, people will spot disease long before they feel sick—as Larry did. They will regard the
doctor as more consultant than oracle.

Not everyone sees this potential revolution as a good one. Do
people really want or need to know this much about themselves? Is
such a preoccupation with health even *healthy?* What if swimming
in oceans of biodata causes more harm than good?

"Frankly, I'd rather go river rafting," says Dr. H. Gilbert Welch,
a professor of medicine at the Dartmouth Institute for Health Policy and Clinical Practice and the author of *Overdiagnosed: Making
People Sick in the Pursuit of Health.* "Data is not information. Information is not knowledge. And knowledge is certainly not wisdom."
Welch believes that individuals who monitor themselves as closely
as Larry does are pretty much guaranteed to find something
"wrong." Contradictory as it sounds, he says abnormality is *normal.*

"It brings to mind the fad a few years ago with getting full-body
CT scans," Welch says. "Something like eighty percent of those
who did it found something abnormal about themselves. The essence of life is variability. Constant monitoring is a recipe for all
of us to be judged 'sick.' Judging ourselves sick, we seek intervention." And intervention, usually with drugs or surgery, he warns, is
never risk-free. Humbler medical practitioners, aware of the sor-
did history of some medical practices (see: bloodletting, lobotomy,
trepanning), weigh the consequences of intervention carefully.
Doing no harm often demands doing nothing. The human body
is, after all, remarkably sturdy and self-healing. As Welch sees it,
"Arming ourselves with more data is guaranteed to unleash a lot
of intervention" on people who are basically healthy.

Not to mention creating an epidemic of anxiety. In other words, the "quantified life" might itself belong to the catalog of affliction, filed under *Looking too closely, hazards of.*

In that sense, the story of Larry Smarr might be less a pioneering saga than a cautionary tale.

Larry's journey started with that most American of preoccupations, losing weight. Larry doesn't update the photo each time he renews his California driver's license, preferring to keep, as a reminder, the one taken soon after his arrival at UCSD twelve years ago with his wife, Janet. It shows a fifty-one-year-old Larry, one with more and longer hair, a wide, round face, and an ample second chin. Call him Jolly Larry. He had just arrived from Illinois, a place he now refers to as "the epicenter of the obesity epidemic," and he had a girth to match his oversize professional reputation. (Deepfried, sugarcoated pastries were a particular favorite of his back then.) Arriving in La Jolla, Jolly Larry found himself surrounded by jogging, hiking, biking, surfing, organic-vegetable-eating superhumans. It was enough to shame him into action. If he was going to fit in on this sunny new campus, he would have to shape up.

So Jolly Larry started working out, reading diet books, and stepping on the scale every day. At first his charts were disappointing. Like countless strivers before him, he dropped some weight, but not much, and it kept wanting to come back. Three or four popular books on weight loss left him mostly confused, but they did convey a central truth: losing weight was only 20 percent about exercise. The other 80 percent was about what he put in his mouth. What triggered his breakthrough was the advice of Barry Sears, the biochemist who created the Zone Diet, which pressed Larry's buttons precisely. Sears proposed that to diet more effectively, one needed to *know* more. Larry decided to study up on his body chemistry.

Few people in history have been better positioned to act on such advice. Larry had begun his professional life as an astrophysicist, trying to unravel the core puzzles of the universe. In 1975, when he was working toward his doctorate at the University of Texas, one of his advisers suggested that he get a top-secret government security clearance: behind the walls of America's nuclear weapons program were not only some of the nation's premier physicists, but also the world's first supercomputers, hundreds of times faster than anything available on any college campus. Larry got his clearance, and in the following years, while working as a fellow at Princeton and at Harvard, he would disappear during summers behind the classified walls of the Lawrence Livermore National Laboratory, in the San Francisco Bay Area. There he would work sixteen-hour shifts on some of the most difficult problems in his field—but with a crucial difference. Working with a computer at one of his universities, Larry might set it a task to compute overnight. He would go home, and when he returned the next morning, the task would be nearing completion. Working with the new Cray supercomputer at Livermore, he could get the same result in a minute and a half.

When he'd return to his university posts in the fall and rejoin his colleagues working at a comparative snail's pace, he'd tell them, "You know, guys, we could be using supercomputers to solve the laws of physics instead of trying to do these closed-form static solutions that you do." They would look at him as if he was crazy. "What are you talking about?" they'd ask. "That can't be done." To them, it seemed impossible. The supercomputer enabled not just faster work but a different style and language of experimentation. But when he tried to explain it to his colleagues, who were still working mostly with pencils and paper, they scratched their heads. "It was like I was living in two different worlds," Larry says.

When one of the first Cray computers outside of secret nuclear programs was set up in Munich, Larry started spending his summers there. "And in about eighty-two, we were at a beer garden and it was probably my second glass of beer, and I was being hosted by a German astrophysicist, world-class," Larry recalls. "He asks, 'Tell me something. My father helped build the trains Germany relied on during the war. And here in our occupied country, you guys, you Americans, come over here and mooch off of our supercomputers because you don't have the wit to put them in your universities where people can get access to them. Have I got that right?' And I said, 'Pretty much.' And he asks, 'How did you guys win the war?'"

Larry brought that question home with him to his perch at the University of Illinois. There, in 1983, he helped draft "The Black Proposal," an unusually concise recommendation (in a black cover) for a $55 million National Science Foundation supercomputer center. When it was funded, along with four other

NSF centers, Larry and others argued for using the protocols of the military's ARPANET (the precursor to the Internet) to link the centers, so that civilian researchers across the nation could use the fastest computers in America for basic research. The linking proposal was controversial, not only because it took on the cult of secrecy surrounding the most advanced computers in America, but because it specifically recommended that the NSF include only computer networks using TCP/IP, a universal computer protocol designed to facilitate not secrecy but collaboration. TCP/IP allowed different kinds of computers to exchange data seamlessly. At the time, the large computer companies—DEC, IBM, General Electric, and so on—preferred a market model in which manufacturers competed to create large fiefdoms, networks that used only their own machines. By adopting Larry's proposal, the NSF enabled computer networks to plug into the system, a critical step toward today's Internet.

By the time, years later, that Larry heeded Barry Sears's suggestion to learn more about his body's chemistry, Larry had at his disposal at UCSD a supercomputer with a capacity many times greater than that of any he'd worked on at Livermore. His research interests had shifted from astrophysics to the impact computers were having on all kinds of fields, including medicine. Calit2 already had numerous grants to study "digitally enabled genomic medicine," so in 2010 Larry signed himself up as a test subject. As his personal quest to lose weight evolved into an effort to understand human biochemistry, his own body became the equivalent of the coral-reef tank he'd once kept in his living room.

Larry had already radically changed his diet, breaking his intake into subcategories, aiming for a caloric split of 40 percent low-glycemic carbohydrates, 30 percent lean protein, and 30 percent omega 3–enriched fat. His meal portions were about half of ordinary restaurant portions. Following what was essentially Barry Sears's Zone Diet, Larry had lost a pound every ten weeks, dropping twenty pounds in four years.

Most people would have been happy with that. But his dieting taught Larry something. If he wanted good health, he could not simply trust how he felt and wing it. If he wanted to understand what was happening in his body, he had to examine the data. And despite his weight loss, the data were now telling him something that didn't seem to make sense. By his calculations, the pounds should still have been falling off, but they weren't.

According to his measurements, he had doubled his strength and tripled the number of steps he took each day. His REM periods, the most valuable periods of sleep, accounted for more than half the time he spent in the sack—twice the typical proportion for a man of his age. His weight was steady. But Larry wanted to know more. He had been getting blood tests once or twice a year as part of his normal health maintenance, but by the end of 2010 he was sending off blood samples more often and graphing dozens of markers, which enabled him to at least better define the mystery. The Zone Diet is designed to reduce inflammation, and because he followed it faithfully, Larry expected his blood-test inflammation score to be low. But the C-reactive protein (CRP), which rises in response to inflammation, was high.

"I had discovered that my body is chronically inflamed—just the opposite of what I expected!" he wrote in an account of his project published last year in a special issue of *Strategic News Service*, a computer/telecommunications newsletter. (The article was prefaced by an enthusiastic note from the publisher, Mark R. Anderson, who said that it "may be the most important Special Letter we have ever published. For many of you reading it, it may also save your lives, or extend them.") Larry wrote: "Even more intriguing: after I had been tracking my CRP for two years, I noticed that it had suddenly more than doubled in less than a year. Troubled, I showed my graphs to my doctors and suggested that something bad was about to happen."

Here you should try to imagine the average physician's reaction when a patient, outwardly healthy, arrives with detailed graphs of his body chemistry, concerned that something evil is stalking his insides.

"Do you have a symptom?" Larry was asked.

"No," he answered. "I feel fine."

He was assured that charts like his were "academic" and not useful for clinical practice. The doctors told him to come back if and when he found something actually wrong with *him,* as opposed to finding anomalies in his charts.

I ask Larry a question his doctors might have been too polite to ask: "Are you a hypochondriac?"

"A hypochondriac is someone who imagines that they have things that are wrong with them and worries about that," he says. "I am the opposite of a hypochondriac. I don't make any assumptions about what might be right or wrong with me, and I don't imagine it. I measure it."

Larry was beginning to have serious doubts about the way medicine is practiced in this country. "Here's the way I look at it; the average American has something like two twenty-minute visits a year with a doctor," he explains. "So you have forty minutes a year that that doctor is going to help you make good decisions. You have five hundred thousand minutes a year on your own, and every one of those, you are making decisions. So we're already in a situation where you are in charge of your ship—your body—and you are making a lot of pretty horrible decisions, or else two-thirds of the United States' citizens wouldn't be overweight or obese. You wouldn't have the CDC saying that forty-two percent of Americans may be obese by 2030, and a third of all Americans may develop diabetes by 2050. That's the result of a lot of bad decisions that people are individually making on their own."

A few weeks after his doctors dismissed his graphs as "academic," Larry felt a severe pain in the left side of his abdomen. At his doctor's office, he was diagnosed with an acute bout of diverticulitis, an intestinal disease caused by inflammation. He was put on a ten-day antibiotic program to treat the ailment. To Larry, this perfectly illustrated the problem. Doctors were ready, eager, and well equipped to address a clinical symptom but unwilling to wade with him into his charts, which, although undeniably abstract, had foretold the problem! It was at this point that Larry decided to take over his own health care.

He asked to see the written report from his last colonoscopy and underwent another. He began testing his stool, recognizing that all of us are, in fact, "superorganisms," that our gastrointestinal, or GI, tracts are a collaboration between human digestive cells and the trillions of bacteria that line our intestines. The stool samples provided detailed charts of the workings of these microorganisms, which is what Larry means when he calls his poop "data-rich." He was learning more about the biochemistry of his own body than any patient had ever known, and the numbers continued to add up in an alarming way. They suggested that he was suffering not from diverticulitis but from some kind of inflamed-bowel disease.

He then went looking for an expert to help him interpret the data. He didn't have to look far: Dr. William J. Sandborn had recently left the Mayo Clinic to take over the GI division of UCSD's School of Medicine.

"I think he felt like he wasn't really being taken seriously," says Sandborn. "So he came over and we looked, and we ended up finding some degree of inflammation that was pointing in the direction of Crohn's disease, but he wasn't really having many symptoms. So the question then became: Is this some kind of early subclinical Crohn's disease? Should we even go as far as treating it, or just wait?"

Larry's impressive quest to fine-tune his body had led him to this: an early diagnosis of Crohn's disease, an incurable condition. It isn't fatal, but it has a long list of uncomfortable and sometimes painful symptoms that tend to flare up from time to time; they center around the GI tract but may include eye inflammation, swollen joints, fever, fatigue, and others. Apart from that one episode of abdominal pain, Larry was still feeling fine. But the graphs showed, and his new doctor more or less confirmed, that he was sick.

And that part about its being incurable? Let's just say that in Larry, Crohn's disease has encountered a very dedicated adversary.

If past thinkers leaned heavily on the steam engine as an all-purpose analogy—for example, contents under pressure will explode (think Marx's ideas on revolution or Freud's about repressed desire)—today we prefer our metaphors to be electronic. We talk about neural "circuitry," about "processing" information, or about how genes "encode" our physical essence. In this worldview, our bodies are computers, and DNA functions as our basic program, our "operating system."

This is certainly how Larry, the computer scientist, talks about the human body. In this context, all of human history can be seen as a progression from a world that was data-poor to one that is data-rich. Starting with those early summers working in secrecy at Livermore, Larry has witnessed firsthand the exponential progress of computing power posited by Moore's Law, which states that the computer-chip transistor count should double roughly every two years. So when Larry talks about the potential for computers to help us understand our bodies, he isn't talking about their showing us more isolated details about an unfathomably complex system; he's talking about knowing *everything*.

"We are going to know—once you know each of your cells' six billion genome bases, with all the imaging down to the micron level, and when you know every damn gene and every bacterium —at a certain point, there is no more data to know," he says. "So certainly by 2030, there is not going to be that much more to learn . . . I mean, you are going to get the wiring diagram, basically." Once they are armed with the wiring diagram, Larry sees no reason why individuals cannot maintain their health the way modern car owners maintain their automobiles.

Larry actually concedes the point made by Dartmouth's Welch —that presented with enough data, pretty much everyone is going to find something wrong with them. He just disputes that this would be a bad thing. "All of us do have something beginning to go wrong, but then, so do our automobiles," Larry says. "In today's world of automobile preventive maintenance, we don't wait for our cars to break down and then go to the 'car doctor.' Every ten thousand or twenty thousand miles, we go in and get an exhaustive look at all the key variables since the last check. If they find something wrong with my car—which will be different from what they find about yours—then they take appropriate action and I go back to driving a 'healthy' car. Occasionally, something is discovered that indicates a bunch of cars need to be called in and get a certain item replaced. I can imagine that occasionally, as a new DNA segment is related to some disease, people with that DNA signature will be called in for 'preventive maintenance.'"

If Larry is right, then our descendants may view early-twenty-first-century medical practices, which we consider a triumph of reason over superstition, in the same way we now view eighteenth- and nineteenth-century folk remedies. A particularly likely candidate for scorn in an age of "quantified" health care is our one-pill-fits-all approach to prescription drugs. In his book *The Creative Destruction of Medicine,* the physician-author Eric Topol cites such dosing as an example of medicine that is "population-based" rather than "patient-centered." He notes the widespread use of statins to lower LDL cholesterol, a factor in heart disease. Topol doesn't deny the cholesterol-lowering effect of these drugs, but he argues that double-blind testing also shows that this effect benefits only a tiny fraction of those treated. One of the most effective statins, Crestor, has been found to reduce the incidence of stroke, heart attack, or death from 4 percent of patients in the placebo group to 2 percent of the group taking the statin. And yet these drugs are widely administered to patients considered at risk. Topol writes: "Instead of identifying the 1 person or 2 people out of every 100 who would benefit, the whole population with the criteria that were tested is deemed treatable . . . What constitutes evidence-based medicine today is what is good for a large population, not for any particular individual."

Pharmaceutical companies don't mind. And as long as the harmful side effects are within acceptable limits, the Food and Drug Administration doesn't mind, either. Some patients will be helped. All of them will be buying the pills, and all will be subjected to follow-up tests, some of them uncomfortable and most of them unnecessary. What if there were a way, Topol asks, of knowing, before prescribing the drug, which 2 percent would be most likely to benefit from it? In an observation that Larry wholeheartedly endorses, Topol writes: "Fortunately, our ability to get just that information is rapidly emerging, [and we are] beginning an era characterized by the right drug, the right dose, and the right screen for the right patients, with the right doctor, at the right cost."

Getting there will mean essentially dismantling the health-care industry as we know it. (Thus the *creative destruction* of Topol's title.) Or, as Larry puts it: "A lot of enormously wealthy, established, powerful institutions in our society are going to be destroyed." And why not? Over the past twenty years, computers have been toppling and rebuilding industries one by one, from retail sales (Walmart and Amazon), to banking (ATMs and online services), to finance (high-speed online investing), to entertainment (web streaming, downloads, YouTube, and so on), to publishing (e-books and news aggregators). We're just babes in this new digital era, and it will eventually upend almost every field of human endeavor.

Larry sees medicine as a stubborn holdout. Current efforts to reform the system—for instance, the Obama administration's initiative to digitize all health records by 2014—are just toes in the water. Medicine has barely begun to take advantage of the million-fold increase in the amount of data available for the diagnosis and treatment of disease. Take the standard annual physical, with its weigh-in, blood-pressure check, and handful of numbers gleaned from select tests performed on a blood sample. To Larry, these data points give your doctor little more than a "cartoon" image of your body. Now imagine peering at the same image drawn from a galaxy of *billions* of data points. The cartoon becomes a high-definition, 3-D picture, with every system and organ in the body measured and mapped in real time.

Indeed, a very early prototype of this kind of high-definition image already exists at Calit2. It is, of course, of Larry.

Inside a "cave" fashioned from large HD screens (each with dual rear projectors) and linked to eighteen gaming PCs to create a graphics supercomputer, Larry and I step *into* a stunning image assembled from an MRI scan of his torso. The room, the size of a walk-in closet, is lined with giant screens, front, sides, and back. More screens angle from these walls toward a floor that is illuminated from above. Two curved, waist-high metal railings offer support, because viewers at the center of this visual world can easily lose their balance. A sensor strapped to your forehead tells the computer where you are looking, so as you turn your head it smoothly blends the images on the screens to create a seamless 360-degree alternative world. (This is clearly the future of video games and cinema.) I had to lean on the metal bars to remind myself I was not someplace else. Once we were in position, Jürgen P. Schulze, a Calit2 research scientist, punched up a display of Larry's own coiled, sixty-three-year-old entrails. I felt as if I could reach out and touch the wrinkled contours of his intestines and arteries.

Larry's inner ten-year-old rejoices. "Look!" he says, lifting and opening his hands. "This is me!"

He points to the source of his health concerns, the precise six-inch stretch of his sigmoid colon that is visibly distorted and inflamed. This is Larry's discovery and his enemy.

I note that the display breaks new ground in the annals of self-disclosure: Larry is literally turning himself inside out for a journalist. He does worry a little about making public such intimate details, but this openness is part of how he believes medicine ought to be—and ultimately *will* be—practiced. The current consensus that medical records should be strictly private, subject to the scrutiny of only doctor and patient, will be yet another casualty if Larry's health-care vision comes to pass. "A different way to organize society is to say it is human-focused, human-centered, patient-centered, and that there are no legal or financial repercussions from sharing data," he says. "There is a huge societal benefit from sharing the data, getting it out from the firewalls, letting software look across millions of these things."

The way the system works now, when a technician examines the MRI of a patient's abdomen, in two dimensions, on a single screen, she compares and contrasts it with perhaps thousands or even tens of thousands of other images she has seen. She then writes a report to the physician explaining, on the basis of her memory and experience, what is normal or abnormal in what she sees.

But "software can go in, volumetrically, over, say, a *million* different abdomens," says Larry, gesturing at the image of his own innards, "and come up with exquisite distribution functions of how things are arranged, what is abnormal or normal, on every little thing in there. In my case, what I have found is inflammation. Unaddressed, it may lead to structural damage and maybe eventually surgery, cutting that part out. So I am going to have another MRI in three months, and that will tell me whether the things I am doing have made it better, or if it is the same, or has gotten worse."

It's that sense of control that appeals to Larry as much as anything.

"The way we do things now," he says, "the technician will examine it and write up a report, which goes to my doctor, and then he explains it all to me. So I am *disembodied*. Patients are completely severed from having any relationship with their body. You are helpless."

Shedding that sense of disembodiment and helplessness is, in theory, one of the most attractive features of Larry Smarr's quantified self. Individuals will understand their own bodies and take care of themselves; doctors will merely assist with the maintenance and fine-tuning. With that sense of personal ownership established, Larry believes, the average American won't continue to drink 500 cans of soda a year, or ingest some 60 pounds of high-fructose corn syrup. After all, educational campaigns about cigarettes have helped lower the share of smokers in America to below 20 percent. If we made such inroads into the obesity epidemic, Larry says, "we would have a national celebration."

For his part, Larry is no longer disembodied. He has had key snippets of his DNA sequenced and will have the whole thing completely sequenced by the end of this year. In just what he has seen so far, he has discovered telltale markers linked with late-onset Crohn's disease. He has developed his own theory of the disease, based on his reading of the most recent medical literature and his growing perception of himself as a superorganism. In a nutshell, he suspects that some of the essential bacteria that should line the walls of his intestine at the point where it is inflamed have been killed off, probably by some antibiotic regimen he underwent years ago. So he has begun charting, through stool samples, the bewilderingly complex microbial ecology of his intestines.

He showed me a detailed analysis of one such sample on his computer, drawing my attention to the word "firmicute." "So, what the hell is a firmicute?" he asks rhetorically. "And in particular, it is in these two groups, *Clostridium leptum* and *Clostridium coccoides*. So I go back, and I go, 'Clostri-Clostri-Clostri, that rings a bell. I had it in my last stool measurement.'" He pulls up an older chart on his screen. "Here is my stool measurement from January 1, 2012. And here are my bacteria. *Lactobacillus* and *Bifidobacteria:* that is what you get in, like, a yogurt and stuff like that, right? *Clostridiums:* you can have them from zero to four-plus. Four-plus is what they should be. And you can see I am deficient here on a number of them," he says, pointing to low numbers on the chart. "So then I went back over time and got them plotted, and they never were above two, and now they are collapsed down to one. So it looks like I am losing. So what do *Clostridia* do? Because I am missing them—I am missing that service."

You may note the *Alice in Wonderland* quality of all this. Every question Larry seeks to answer raises new questions, every door he opens leads to a level of more bewildering complexity. One could easily conclude that these levels never bottom out, that the intricacy of the human body, composed of its trillions of cells— each dancing to the tune of a genetic program but also subject to random intersections with outside forces such as radiation, chemicals, and physical accidents—is for all practical purposes infinite, and hence permanently beyond our full comprehension. But Larry, with his astrophysics background, is utterly undaunted by complexity. This is the gift of the computer age: things once considered too numerous to count can now be counted. And Larry believes that questions about how the human body functions are ultimately finite.

In his own case, Larry has zeroed in on what he believes is the specific missing bacterial component behind the immune-system malfunction causing his bowel inflammation. He's begun a regimen of supplements to replace that component. If it doesn't work, he'll devise a new plan. He isn't aiming for immortality—not yet —although as far as he is concerned, it's not out of the question. As we develop our ability to replace broken-down body parts with bioengineered organs, and as we work toward a complete understanding of human systems and biochemistry . . . why not?

Reflecting on Larry's vision of a patient-centric, computer-assisted world of medical care, Dr. Welch allows: "I can conceive of this happening. But is this the model we want for good health? What does it mean to be healthy? Is it something we learn from a machine? Is it the absence of abnormality? Health is a state of mind. I don't think constantly monitoring yourself is the right path to that state of mind. Data alone is not the answer. We went through all of this with the Human Genome Project. You heard it then: if we could just get all of this data, all of our problems would be solved. It turned out that the predictive power of mapping the genome wasn't all that great, because there are other factors at play: the environment, behavior, and chance. Randomness has a lot to do with it."

And these are not the only reasons to be skeptical of Larry's vision. Researchers will certainly continue to map the human body in ever-greater detail, enabling doctors to spot emerging illness earlier and to design drug treatments with far more precision. But in the end, how many people will want to track their bodily functions the way Larry does, even if software greatly simplifies the task? Larry says the amount of time he has spent monitoring and studying himself has grown a lot, but that it still adds up to less time each day than most Americans spend watching television. But even if that time is radically reduced by software, how many of us, understanding that our decrypted genome may reveal terrible news about our future—Alzheimer's, crippling neuromuscular diseases, schizophrenia, and so on—will even want to know?

When I ask Larry this question, he frowns and says, "I can't understand that." The very idea stumps him. To him, *not want-ing to know* something—even bad news—just doesn't compute. His whole life is about finding out. He's a scientist to his core.

"I hear it a lot, but I don't understand it. Because whatever it is, if you suspect that you are going to have, say, Alzheimer's within five years or ten years, then that should focus your mind on what it is you want to accomplish in the days that you have left." Then, after a moment more thought, he adds, "And if you don't know, those days are going to just slide by, in which you could have done something that you always meant to do."

He knows that the way he lives and works might seem eccen-tric or even a little crazy to others. "Most of my life, people have thought I was crazy at any given point," he says. "Maybe being crazy simply means you are clear-sighted and you are looking at the fact that you are in a period of rapid change. I see the world as it will be, and of course, that is a different world than the one we live in now."

Larry is in a hurry to get there. He sees himself ten years down the road as someone healthy and active and strong, instead of someone struggling to manage the increasingly uncomfortable and debilitating effects of Crohn's. As he makes his way down the supplements aisle of his Whole Foods Market, looking for a very specific assortment of probiotics with which to mix his remedial cocktail, he's not just trying to save himself. He's trying to save you.

KEVIN DUTTON

The Wisdom of Psychopaths

FROM *Scientific American*

TRAITS THAT ARE common among psychopathic serial killers—a grandiose sense of self-worth, persuasiveness, superficial charm, ruthlessness, lack of remorse, and manipulation of others—are also shared by politicians and world leaders. Individuals, in other words, running not from the police but for office. Such a profile allows those who present with these traits to do what they like when they like, completely unfazed by the social, moral, or legal consequences of their actions.

If you are born under the right star, for example, and have power over the human mind as the moon over the sea, you might order the genocide of 100,000 Kurds and shuffle to the gallows with such arcane recalcitrance as to elicit, from even your harshest detractors, perverse, unspoken deference.

"Do not be afraid, doctor," said Saddam Hussein on the scaffold, moments before his execution. "This is for men."

If you are violent and cunning, like the real-life "Hannibal Lecter," Robert Maudsley, you might take a fellow inmate hostage, smash his skull in, and sample his brains with a spoon as nonchalantly as if you were downing a soft-boiled egg. (Maudsley, by the way, has been cooped up in solitary confinement for the past thirty years in a bulletproof cage in the basement of Wakefield Prison in England.)

Or if you are a brilliant neurosurgeon, ruthlessly cool and focused under pressure, you might, like the man I'll call Dr. Geraghty, try your luck on a completely different playing field: at the remote outposts of twenty-first-century medicine, where risk blows in on 100-mile-per-hour winds and the oxygen of deliberation is thin. "I have no compassion for those whom I operate on," he told me. "That is a luxury I simply cannot afford. In the theater I am reborn: as a cold, heartless machine, totally at one with scalpel, drill, and saw. When you're cutting loose and cheating death high above the snowline of the brain, feelings aren't fit for purpose. Emotion is entropy—and seriously bad for business. I've hunted it down to extinction over the years."

Geraghty is one of the UK's top neurosurgeons—and although on one level his words send a chill down the spine, on another they make perfect sense. Deep in the ghettoes of some of the brain's most dangerous neighborhoods, the psychopath is glimpsed as a lone and merciless predator, a solitary species of transient, deadly allure. No sooner is the word out than images of serial killers, rapists, and mad, reclusive bombers come stalking down the sidewalks of our minds.

But what if I were to paint you a different picture? What if I were to tell you that the arsonist who burns your house down might also, in a parallel universe, be the hero most likely to brave the flaming timbers of a crumbling, blazing building to seek out, and drag out, your loved ones? Or that the kid with a knife in the shadows at the back of the movie theater might well, in years to come, be wielding a rather different kind of knife at the back of a rather different kind of theater?

Claims like these are admittedly hard to believe. But they're true. Psychopaths are fearless, confident, charismatic, ruthless, and focused. Yet contrary to popular belief, they are not necessarily violent. Far from its being an open-and-shut case—you're either a psychopath or you're not—there are, instead, inner and outer zones of the disorder: a bit like the fare zones on a subway map. There is a spectrum of psychopathy along which each of us has our place, with only a small minority of A-listers resident in the "inner city."

Think of psychopathic traits as the dials on a studio mixing deck. If you turn all of them to max, you'll have a soundtrack that's no use to anyone. But if the soundtrack is graded, and some are up higher than others—such as fearlessness, focus, lack of empathy, and mental toughness, for example—you may well have a surgeon who's a cut above the rest.

Of course, surgery is just one instance where psychopathic "talent" may prove advantageous. There are others. In 2009, for instance, I decided to perform my own research to determine whether, if psychopaths were really better at decoding vulnerability (as had been found in some studies), there could be applications. There had to be ways in which, rather than being a drain on society, this ability actually conferred some benefit. And there had to be ways to study it.

Enlightenment dawned when I met a friend at the airport. We all get a bit paranoid going through customs, I mused. Even when we're perfectly innocent. But imagine what it would feel like if we did have something to hide—and if an airport security officer were particularly good at picking up on that feeling?

To find out, I decided to conduct an experiment. Thirty undergraduate students took part: half of them high on the Self-Report Psychopathy Scale, and half of them low. There were also five "associates." The students' job was easy. They had to sit in a classroom and observe the associates' movements as they entered through one door and exited through another, traversing en route a small, elevated stage. But there was a catch. They also had to note who was "guilty": Which one of the five was concealing a scarlet handkerchief?

To raise the stakes and give the observers something to "go on," the associate with the handkerchief was handed £100. If the jury decided that he was the guilty party—if, when the votes were counted, he came out on top—then he had to hand it back. If, on the other hand, he got away with it and the finger of suspicion fell heavier on one of the others, he would stand to be rewarded. He would instead get to keep the £100.

Which of the students would make the better "customs officers"? Would the psychopaths' predatory instincts prove reliable? Or would their nose for vulnerability let them down?

More than 70 percent of those who scored high on the Self-Report Psychopathy Scale correctly picked out the handkerchief-smuggling associate, compared with just 30 percent of the low scorers. Zeroing in on weakness may well be part of a serial killer's tool kit. But it may also come in handy at the airport.

Trolleyology

Joshua Greene, a psychologist at Harvard University, has observed how psychopaths unscramble moral dilemmas. As I described in

my 2011 book, *Split-Second Persuasion,* he has stumbled on something interesting. Far from being uniform, empathy is schizophre-
nic. There are two distinct varieties: hot and cold.

Consider, for example, the following conundrum (Case 1), first
proposed by the late philosopher Philippa Foot:

A railway trolley is hurtling down a track. In its path are five
people who are trapped on the line and cannot escape. Fortunately, you can flip a switch that will divert the trolley down a fork
in the track away from the five people—but at a price. There is
another person trapped down that fork, and the trolley will kill
him or her instead. Should you hit the switch?

Most of us experience little difficulty when deciding what to
do in this situation. Although the prospect of flipping the switch
isn't exactly a nice one, the utilitarian option—killing just the
one person instead of five—represents the "least worst choice."
Right?

Now consider the following variation (Case 2), proposed by the
philosopher Judith Jarvis Thomson:

As before, a railway trolley is speeding out of control down a
track toward five people. But this time you are standing behind a
very large stranger on a footbridge above the tracks. The only way
to save the five people is to heave the stranger over. He will fall to
a certain death. But his considerable girth will block the trolley,
saving five lives. Question: Should you push him?

Here you might say we're faced with a "real" dilemma. Although
the score in lives is precisely the same as in the first example (five
to one), playing the game makes us a little more circumspect and
jittery. But why?

Greene believes he has the answer. It has to do with different
climatic regions in the brain.

Case 1, he proposes, is what we might call an impersonal moral
dilemma and involves those areas of the brain, the prefrontal cortex and posterior parietal cortex (in particular, the anterior para-
cingulate cortex, the temporal pole, and the superior temporal
sulcus), principally implicated in our objective experience of cold
empathy: in reasoning and rational thought.

Case 2, on the other hand, is what we might call a personal
moral dilemma. It hammers on the door of the brain's emotion
center, known as the amygdala—the circuit of hot empathy.

Just like most normal members of the population, psychopaths make pretty short work of the dilemma presented in Case 1. Yet— and this is where the plot thickens—quite unlike normal people, they also make pretty short work of Case 2. Psychopaths, without batting an eye, are perfectly happy to chuck the fat guy over the side.

To compound matters further, this difference in behavior is mirrored rather distinctly in the brain. The pattern of neural activation in both psychopaths and normal people is well matched on the presentation of impersonal moral dilemmas—but dramatically diverges when things get a bit more personal.

Imagine that I were to pop you into a functional MRI machine and then present you with the two dilemmas. What would I observe as you went about negotiating their moral minefields? Just around the time that the nature of the dilemma crossed the border from impersonal to personal, I would see your amygdala and related brain circuits—your medial orbitofrontal cortex, for example—light up like a pinball machine. I would witness the moment, in other words, that emotion puts its money in the slot.

But in a psychopath, I would see only darkness. The cavernous neural casino would be boarded up and derelict—the crossing from impersonal to personal would pass without any incident.

The Psychopath Mix

The question of what it takes to succeed in a given profession, to deliver the goods and get the job done, is not all that difficult when it comes down to it. Alongside the dedicated skill set necessary to perform one's specific duties—in law, in business, in whatever field of endeavor you care to mention—exists a selection of traits that code for high achievement.

In 2005 Belinda Board and Katarina Fritzon, then at the University of Surrey in England, conducted a survey to find out precisely what it was that made business leaders tick. What, they wanted to know, were the key facets of personality that separated those who turn left when boarding an airplane from those who turn right?

Board and Fritzon took three groups—business managers, psychiatric patients, and hospitalized criminals (those who were psychopathic and those suffering from other psychiatric illnesses) —and compared how they fared on a psychological profiling test.

Their analysis revealed that a number of psychopathic attributes were actually more common in business leaders than in so-called disturbed criminals—attributes such as superficial charm, egocentricity, persuasiveness, lack of empathy, independence, and focus. The main difference between the groups was in the more "antisocial" aspects of the syndrome: the criminals' lawbreaking, physical aggression, and impulsivity dials (to return to our analogy of earlier) were cranked up higher.

Other studies seem to confirm the "mixing deck" picture: that the border between functional and dysfunctional psychopathy depends not on the presence of psychopathic attributes per se but rather on their levels and the way they are combined. Mehmet Mahmut and his colleagues at Macquarie University in Sydney, Australia, have recently shown that patterns of brain dysfunction (specifically, patterns in orbitofrontal-cortex functioning, the area of the brain that regulates the input of the emotions in decision making) observed in both criminal and noncriminal psychopaths exhibit dimensional rather than discrete differences. This, Mahmut suggests, means that the two groups should not be viewed as qualitatively distinct populations but rather as occupying different positions on the same continuum.

In a similar (if less high-tech) vein, I asked a class of first-year undergraduates to imagine they were managers in a job placement company. "Ruthless, fearless, charming, amoral, and focused," I told them. "Suppose you had a client with that kind of profile. To which line of work do you think they might be suited?"

Their answers couldn't have been more insightful: CEO, spy, surgeon, politician, the military . . . they all popped up in the mix. Among serial killer, assassin, and bank robber.

"Intellectual ability on its own is just an elegant way of finishing second," one successful CEO told me. "Remember, they don't call it a greasy pole for nothing. The road to the top is hard. But it's easier to climb if you lever yourself up on others. Easier still if they think something's in it for them."

Jon Moulton, one of London's most successful venture capitalists, agrees. In a recent interview with the *Financial Times*, he lists determination, curiosity, and insensitivity as his three most valuable character traits.

No prizes for guessing the first two. But insensitivity? The great thing about insensitivity, Moulton explains, is that "it lets you sleep when others can't."

Contributors' Notes

Other Notable Science and
Nature Writing of 2012

Contributors' Notes

Natalie Angier, a science columnist for the *New York Times,* is the author of *Woman: An Intimate Geography* and *The Canon: A Whirligig Tour of the Beautiful Basics of Science,* among other books. She decided to tackle the topic of infinity when her sixteen-year-old daughter came home with an amusing calculus assignment. Imagine a battle between two superheroes, the teacher said, named Captain Zero and Infinitus. Who would win and under what circumstances? "It was a lesson about limits," her daughter said. The limits of infinity? Irresistible.

Rick Bass is the author of thirty-one books of fiction and nonfiction. He lives in northwest Montana's Yaak Valley, where he is a board member of the Yaak Valley Forest Council and Round River Conservation Studies. His stories and essays have been anthologized in *The Best American Short Stories, The Best Spiritual Writing,* and *The Best American Travel Writing.*

Mark Bowden is an author and longtime journalist, best known for his book *Black Hawk Down.* Once upon a time he was a science writer for the *Philadelphia Inquirer.*

Gareth Cook is a Pulitzer Prize–winning magazine journalist and a regular contributor to NewYorker.com. His work has appeared in the *New York Times Magazine, Wired, Scientific American, Washington Monthly,* the *Boston Globe,* and elsewhere. He is the editor of *Scientific American*'s Mind Matters neuroscience blog and the series editor of *The Best American Infographics.* He was the *Globe*'s science reporter for seven years, one of the founders of the *Globe*'s Sunday Ideas section, and then its editor from 2007 to 2011, when he was named a Sunday columnist. He has also worked for *Washington Monthly, Foreign Policy, U.S. News & World Report,* and the *Boston Phoenix.*

This is the second time his work has appeared in *The Best American Science and Nature Writing.*

David Deutsch, a University of Oxford physicist and the inventor of the concept of universal quantum computers, says he got interested in physics as a child when he rebelled at the claim that no one can understand everything that is understood.

Kevin Dutton is a research psychologist at the Calleva Research Centre for Evolution and Human Sciences, Magdalen College, University of Oxford. He is a member of the Royal Society of Medicine and the Society for the Scientific Study of Psychopathy. Dutton's first book, *Flipnosis: The Art of Split-Second Persuasion,* first published in 2010 and since translated into eighteen languages, documents his quest—from the political genius of Winston Churchill to the malign influence of some of the world's top con artists—for the psychological "DNA" of persuasion. His second book, *The Wisdom of Psychopaths: Lessons in Life from Saints, Spies, and Serial Killers,* was published in 2012. In it Dutton explores the positive side of being a psychopath and discovers firsthand, in a groundbreaking "how to make a psychopath" experiment, what it's like to be one. The effects have since worn off.

Sylvia A. Earle is the former chief scientist of the U.S. National Oceanic and Atmospheric Administration and a National Geographic Explorer-in-Residence, as well as the founder of the Sylvia Earle Alliance, Mission Blue, and Deep Ocean Exploration and Research. Dr. Earle has been called "Her Deepness" by *The New Yorker* and the *New York Times* and has been named a "Living Legend" by the Library of Congress and the first "Hero for the Planet" by *Time.* She has lectured in more than eighty countries, led more than a hundred expeditions—including the first team of women aquanauts—and logged 7,000 hours underwater, with a record solo dive to 1,000 meters and ten saturation dives.

Artur Ekert pioneered entanglement-based cryptography as a graduate student. He is now the director of the Center for Quantum Technologies in Singapore and a professor at Oxford's Mathematical Institute. He is a keen pilot and diver.

Brett Forrest has contributed to *Playboy, Vanity Fair, The Atlantic,* and *National Geographic.* He has lived in Russia, Ukraine, and Brazil.

Keith Gessen was born in Russia in 1975 and emigrated to the United

States with his family in 1981. He is a founding editor of *n+1* and the author of *All the Sad Young Literary Men*, a novel.

Jerome Groopman holds the Dina and Raphael Recanati Chair of Medicine at Harvard Medical School and is chief of experimental medicine at the Beth Israel Deaconess Medical Center. He received his BA from Columbia College summa cum laude and his MD from Columbia College of Physicians and Surgeons in New York. He served his internship and residency in internal medicine at Massachusetts General Hospital and held specialty fellowships in hematology and oncology at the University of California, Los Angeles, and the Dana-Farber/Boston Children's Hospital Cancer Center, Harvard Medical School, in Boston. He serves on many scientific editorial boards and has published more than 180 scientific articles. In 2000 he was elected to the Institute of Medicine of the National Academies. He writes regularly about biology and medicine for lay audiences as a staff writer at *The New Yorker*. In 2011 he coauthored, with Dr. Pamela Hartzband, *Your Medical Mind: How to Decide What Is Right for You.*

Benjamin Hale is the author of the novel *The Evolution of Bruno Littlemore*, which was nominated for the Dylan Thomas Prize and the New York Public Library's Young Lions of Fiction Award. He is a recipient of a Michener-Copernicus Fellowship and the Bard Fiction Prize, and his writing has appeared in *Harper's Magazine, Conjunctions,* the *New York Times, The Millions,* and *Dissent,* among other publications. He teaches at Bard College.

Katherine Harmon is an award-winning freelance writer and contributing editor for *Scientific American*. Her work covers health, biology, food, the environment, and general-interest stories and has appeared in books, magazines, newspapers, and websites, including *Gourmet, Wired,* and *Nature*. Her first book, *Octopus! The Most Mysterious Creature in the Sea,* will be published this fall. She lives in Longmont, Colorado, with her fiancé and their dog. When not writing or editing, she is likely to be running, gardening, or reading about something nerdy. Read more on her website, www.katherineharmon.com.

Elizabeth Kolbert is a staff writer for *The New Yorker* and the author of *Field Notes from a Catastrophe: Man, Nature, and Climate Change*. Her series on global warming, "The Climate of Man," from which the book was adapted, won the American Association for the Advancement of Science's magazine writing award and a National Academies communications award. She is a two-time National Magazine Award winner and has received a Heinz

Award and a Lannan Literary Fellowship. Kolbert lives in Williamstown, Massachusetts.

Alan Lightman is a novelist, essayist, and physicist. He is currently a professor of the practice of the humanities at the Massachusetts Institute of Technology. Lightman's essays and articles have appeared in *Harper's Magazine, The Atlantic, The New Yorker,* and other publications. His novels include *Einstein's Dreams* and *The Diagnosis,* which was a finalist for the National Book Award.

J.B. MacKinnon is a journalist, essayist, and the author, most recently, of *The Once and Future World: Nature As It Was, As It Is, As It Could Be.* His book *Plenty* coined the term "100-mile diet" and is widely recognized as a catalyst of the local food movement. He has lived in the United States, Canada, Spain, and the Dominican Republic.

Stephen Marche is a novelist and a contributing editor at *Esquire.*

Michael Moyer is the special projects editor at *Scientific American,* the leader of the magazine's space and physics coverage, and an award-winning science writer. Before he went into journalism, Moyer studied physics and the philosophical foundations of physics at Columbia University and the University of California, Berkeley, and worked as a researcher with the Nobel Prize–winning Supernova Cosmology Project. He lives in New York's Hudson River Valley with his wife and son.

Michelle Nijhuis writes about science and the environment for *National Geographic, Smithsonian,* the *New York Times,* and many other publications, and she is a longtime contributing editor of *High Country News,* a magazine that covers environmental issues in the American West. Her work has been recognized with several national honors, including two AAAS Science Journalism Awards, and she is the coeditor of *The Science Writers' Handbook: Everything You Need to Know to Pitch, Publish, and Prosper in the Digital Age.* A lapsed biologist, she was once paid to chase tortoises through the Sonoran Desert (the tortoises usually won). She lives with her family in the Columbia River Gorge.

David Owen has been a staff writer for *The New Yorker* since 1991. Before joining *The New Yorker,* he was a contributing editor at *The Atlantic* and, prior to that, a senior writer at *Harper's Magazine.* He is the author of more than a dozen books, including *Copies in Seconds,* about the invention of the Xerox machine; *Green Metropolis,* about the environmental value of urban density; and *The Conundrum: How Scientific Innovation, Increased Efficiency, and Good Intentions Can Make Our Energy and Climate Problems Worse.*

John Pavlus is a filmmaker and writer interested in science, math, design, technology, and other ways in which people make things make sense. His work has appeared in *Scientific American, Wired, Fast Company, Technology Review,* and elsewhere. He creates original short films and documentaries, with partners including NPR, Autodesk, the Howard Hughes Medical Institute, and the *New York Times Magazine,* via his production company, Small Mammal. He lives in Portland, Oregon.

David Quammen is the author of twelve books, including most recently *Spillover: Animal Infections and the Next Human Pandemic,* from which "Out of the Wild" is excerpted. As a contributing writer for *National Geographic,* he travels often, usually to jungles, deserts, savannas, and swamps. In 2012 he received the Stephen Jay Gould Prize from the Society for the Study of Evolution. He lives in Montana with his wife, Betsy Gaines Quammen, and their menagerie.

Nathaniel Rich is the author of two novels, *Odds Against Tomorrow* and *The Mayor's Tongue.* His essays and criticism appear in the *New York Review of Books, Harper's Magazine, Rolling Stone,* and the *New York Times Magazine,* among other publications.

Oliver Sacks is a physician and the author of many books, including *The Man Who Mistook His Wife for a Hat, Musicophilia,* and *Hallucinations.* His book *Awakenings* has inspired a number of dramatic adaptations, including a play by Harold Pinter, an Oscar-nominated film, and a documentary by Bill Morrison with music by Philip Glass. Dr. Sacks is a professor of neurology at the NYU School of Medicine and a visiting professor at Warwick University in the UK.

Robert M. Sapolsky is a professor of biology and neurology at Stanford University and a research associate with the Institute of Primate Research, National Museum of Kenya. A neuroendocrinologist, he focuses his research on the effects of stress. Two of his books, *Why Zebras Don't Get Ulcers: An Updated Guide to Stress, Stress-Related Disease, and Coping* and *The Trouble with Testosterone: And Other Essays on the Biology of the Human Predicament,* were Los Angeles Times Book Club finalists. He has received numerous honors and awards for his work, including a MacArthur Fellowship and an Alfred P. Sloan Fellowship.

Michael Specter has been a staff writer at *The New Yorker* since 1998. He writes often about science, technology, and global public health. Since joining the magazine, he has written several articles about the global AIDS epidemic, as well as about avian influenza, malaria, the world's diminishing freshwater resources, synthetic biology, the attempt to create edible meat in a lab, and the debate over the meaning of our carbon footprint. He has also published many profiles of people, including Lance Armstrong, the ethicist Peter Singer, Sean (P. Diddy) Combs, Manolo Blahnik, and Miuccia Prada. Specter came to *The New Yorker* from the *New York Times,* where he had been a roving foreign correspondent based in Rome. From 1995 to 1998 he served as the *New York Times'* Moscow bureau chief. He came to the *Times* from the *Washington Post,* where from 1985 to 1991 he covered local news, before becoming the national science reporter and later the newspaper's New York bureau chief. Specter has twice received the Global Health Council's annual Excellence in Media Award, first for his 2001 article about AIDS, "India's Plague," and second for his 2004 article "The Devastation," about the ethics of testing HIV vaccines in Africa. He received the 2002 AAAS Science Journalism Award for his 2001 article "Rethinking the Brain," on the scientific basis of how we learn. His most recent book, *Denialism: How Irrational Thinking Hinders Scientific Progress, Harms the Planet, and Threatens Our Lives,* received the 2009 Robert P. Balles Annual Prize in Critical Thinking, presented by the Committee for Skeptical Inquiry. In 2011 Specter won the World Health Organization's Stop TB Partnership Annual Award for Excellence in Reporting for his *New Yorker* article "A Deadly Misdiagnosis," about the dangers of inaccurate TB tests in India, which has the highest rate of tuberculosis in the world. Specter splits his time between Brooklyn and upstate New York.

Steven Weinberg is a professor of physics and astronomy at the University of Texas. His honors include the Nobel Prize in Physics, the National Medal of Science, election to numerous academies, and sixteen honorary doctoral degrees. He has written more than 300 articles on elementary-particle theory, cosmology, and related topics, and twelve books; the latest, *Lake Views: This World and the Universe,* is a collection of his essays in the *New York Review of Books* and other periodicals. Educated at Cornell, Copenhagen, and Princeton, he taught at Columbia, Berkeley, MIT, and Harvard, where he was the Higgins Professor of Physics, before coming to Texas in 1982.

Tim Zimmermann is a correspondent at *Outside* magazine and the author of *The Race.* He writes often about marine mammals and the oceans and recently was an associate producer and cowriter of *Blackfish,* a documentary about killer whales in captivity. Zimmermann's work has appeared in numerous publications, including *Men's Journal* and *Sports Illustrated.* Prior to his work at *Outside,* he was a diplomatic correspondent for *U.S. News & World Report.* More on his work can be found at timzimmermann .com and on Twitter (@Earth_ist). He lives in Washington, D.C., with his wife and two children.

Other Notable Science and Nature Writing of 2012

Selected by Tim Folger

Jill U. Adams
  Chasing Dragons. *Audubon.* July/August.
Isaac Anderson
  The Lord God Bird. *Image.* Issue 72.

Yudhijit Bhattacharjee
  A Week in Stockholm. *Science.* April 6.
Roy Blount Jr.
  Go, Higgs! Block That Boson! *Sports Illustrated.* July 30.
Belle Boggs
  The Art of Waiting. *Orion.* March/April.

Rob Dunn
  The Glory of Leaves. *National Geographic.* October.

Beth Ann Fennelly
  Observations from the Jewel Rooms. *Ecotone.* Fall.
Timothy Ferris
  Sun Struck. *National Geographic.* June.
William Finnegan
  Slow and Steady. *The New Yorker.* January 23.
Douglas Fox
  Witness to an Antarctic Meltdown. *Scientific American.* July.
  The Clouds Are Alive. *Discover.* April.
Susan Freinkel
  In Each Shell a Story. *On Earth.* Summer.

JUSTIN GILLIS
  A Climate Scientist Battles Time and Mortality. *The New York Times*. July 2.
EVAN R. GOLDSTEIN
  The Strange Neuroscience of Immortality. *The Chronicle of Higher Education*.
  July.
ANDREW GRANT
  William Borucki: Planet Hunter. *Discover*. December.
JEROME GROOPMAN
  Sex and the Superbug. *The New Yorker*. October 1.

ROWAN JACOBSEN
  The Gumbo Chronicles. *Outside*. April.
  Boilover. *Outside*. October.
MARK JENKINS
  Last of the Cave People. *National Geographic*. February.

JEFFREY KLUGER
  Fabiola Gianotti, the Discoverer. *Time*. December 19.

MARIANNE LAVELLE
  Good Gas, Bad Gas. *National Geographic*. December.

CHARLES C. MANN
  State of the Species. *Orion*. November/December.
KATHLEEN McAULIFFE
  How Your Cat Is Making You Crazy. *The Atlantic*. March.
BILL McKIBBEN
  A Matter of Degrees. *Orion*. July/August.
  Global Warming's Terrifying New Math. *Rolling Stone*. July 19.
  Scary Monsters. *Orion*. March/April.
ZEEYA MERALI
  Gravity off the Grid. *Discover*. March.
GREGORY MONE
  Frozen. Irradiated. Desolate. Alive? *Discover*. November.
KATHLEEN DEAN MOORE
  Concrete Footing. *Orion*. July/August.
GEORGE MUSSER
  A New Enlightenment. *Scientific American*. November.

BRADFORD PLUMMER
  The Big Crackup. *Audubon*. September/October.

MARY ROACH
  Say Hello to My Little Friend. *Outside*. January.
RON ROSENBAUM
  Richard Clarke on Who Was Behind the Stuxnet Attack. *Smithsonian*. April.